ISBN 978-0-260-12035-9
PIBN 11020252

English
Français
Deutsche
Italiano
Español
Português

www.forgottenbooks.com

Mythology Photography **Fiction**
Fishing Christianity **Art** Cooking
Essays Buddhism Freemasonry
Medicine **Biology** Music **Ancient
Egypt** Evolution Carpentry Physics
Dance Geology **Mathematics** Fitness
Shakespeare **Folklore** Yoga Marketing
Confidence Immortality Biographies
Poetry **Psychology** Witchcraft
Electronics Chemistry History **Law**
Accounting **Philosophy** Anthropology
Alchemy Drama Quantum Mechanics
Atheism Sexual Health **Ancient History**
Entrepreneurship Languages Sport
Paleontology Needlework Islam
Metaphysics Investment Archaeology
Parenting Statistics Criminology
Motivational

PRODUCTS

OF

FREDERICK STEARNS & CO

MANUFACTURING PHARMACISTS.

DETROIT, U.S.A.

CANADIAN LABORATORY
WINDSOR, ONT.
BRANCHES
LONDON, ENG.,   NEW YORK CITY.

Change Sheet No. 5

WE QUOTE the following prices ruling on this date, subject to change without notice. This list supplements our Pharmaceutical Price List "B" and supersedes all previous change sheets. It is subject to the regular discount, except where otherwise specified.

FREDERICK STEARNS & CO.
Manufacturing Pharmacists
Detroit, Mich., U. S. A.
Windsor, Ont.   London, Eng.   New York City

FLUID EXTRACTS

Page.	No.		Per ℔.
4.	4.	Aconite root	$1 50
6.	23.	Angelica root	1 50
6.	28.	Arbor Vitæ	1 50
6.	32.	Arnica root	1 50
6.	34.	Asafetida	1 75
7.	37.	Balm	1 60
7.	38.	Balmony	1 20
8.	47.	Belladonna leaves	1 50
8.	51.	Benzoin comp. for U.S.P. Tincture	1 75
8.	53.	Berberis aquifolium	3 00
9.	59.	Black Alder	1 20
9.	65.	Black Haw	1 75
10.	74.	Blood-root	1 25
10.	77.	Blue Flag	1 25
10.	81.	Broom	1 25
11.	83.	Buchu	2 25
11.	86.	Buchu, Juniper and Pot. acetate	1 90
12.	93.	Buckthorn berries	1 25
12.	94.	Bugleweed	1 20
12.	102.	Calamus	1 50
13.	107.	Canada Snakeroot	1 25

FLUID EXTRACTS
(*Continued*)

FLUID EXTRACTS
(Continued)

SOLID EXTRACTS

POWDERED EXTRACTS

POWDERED EXTRACTS
(*Continued*)

PAGE.	No.		Per oz.
60.	16.	Colchicum root	$0 65
61.	30.	Golden Seal	0 90
61.	32.	Ipecac	3 50
61.	36.	Opium	1 45

SUGAR- AND GELATIN-COATED PILLS

PAGE.	No.		Per 100	Per 500
84.	198.	Cinchonidine, 1 gr	$0 60	$2 85
84.	199.	Cinchonidine, 2 grs. . .	0 85	4 10
84.	200.	Cinchonidine, 3 grs. . .	1 00	4 85
84	201.	Cinchonid. sal., 2 1-2 grs.	1 00	4 85
84.	202.	Cinchonid. sulph., 1-2 gr.	0 40	1 85
84.	203.	Cinchonid. sulph., 1 gr. .	0 50	2 35
84.	204.	Cinchonid. sulph., 2 grs.	0 70	3 35
84.	205.	Cinchonid. sulph., 3 grs.	0 85	4 10
84.	206.	Cinchonid. sulph., 4 grs.	1 10	5 35
84.	207.	Cinchonid. sulph., 5 grs.	1 35	6 60
84.	208.	Cinchonidine Comp. . .	0 60	2 85
84.	209.	Cinchonidine Comp. with Strychnine	0 60	2 5
84.	210.	Cinchonidine and Iron. .	0 60	2 85
85.	211.	Cinchonidine, Iron and Strychnine	0 60	2 85
104.	420.	Morphine acetate, 1-8 gr.	0 40	1 85
104.	421.	Morphine acetate, 1-4 gr.	0 65	3 10
105.	423.	Morph. hydrochlor. 1-6 gr.	0 50	2 35
105.	424.	Morph. hydrochlor. 1-4 gr.	0 65	3 10
	745.	Morph. hydrochlor. 1-2 gr. (added)	1 05	5 10
105.	427.	Morphine sulph., 1-16 gr.	0 25	1 10
105.	428.	Morphine sulph., 1-10 gr.	0 30	1 35
105.	429.	Morphine sulph., 1-8 gr.	0 30	1 35
105.	430.	Morphine sulph., 1-6 gr.	0 35	1 60
105.	431.	Morphine sulph., 1-4 gr.	0 50	2 35
105.	432.	Morphine sulph., 1-2 gr.	0 95	4 60
106.	438.	Night Sweat	0 80	3 85
115.	528.	Pot. permang., 1-8 gr. .	0 20	0 85
115.	529.	Pot. permang., 1-6 gr. .	0 20	0 85
116.	530.	Pot. permang., 1-4 gr. .	0 20	0 85

SUGAR- AND GELATIN-COATED PILLS
(*Continued*)

PAGE.	No.		Per 100	Per 500
116.	531.	Pot. permang., 1-2 gr.	$0 20	$0 85
116.	532.	Pot. permang., 1 gr.	0 30	1 35
116.	533.	Pot. permang., 2 grs.	0 40	1 85
121.	588.	Salicin, 1 gr.	0 45	2 10
121.	589.	Salicin, 2 grs.	0 75	3 60
121.	590.	Salicin, 3 grs.	1 00	4 85
122.	598.	Salol, 2 1-2 grs.	0 45	2 10
122.	599.	Salol, 5 grs.	0 65	3 10
122.	602.	Santonin, 1-2 gr.	0 45	2 10
122.	603.	Santonin, 1 gr.	0 70	3 35
124.	635.	Sumbul Comp. "A"	0 50	2 35
125.	636.	Sumbul Comp. "B"	0 50	2 35

HYPODERMIC TABLETS

PAGE.	No.		Per tube of 25	Per 100 (4 tubes)	Per 100 in 1 bottle
136.	8.	Cocaine hydrochlorate, 1-8 gr.	$0 15	$0 55	$0 45
136.	9.	Cocaine hydrochlorate, 1-4 gr.	0 27	1 00	0 90
137.	25.	Morph. sulph.,1-4 gr.	0 17	0 60	0 50
137.	26.	Morph. sulph.,1-3 gr.	0 22	0 80	0 70
137.	27.	Morph. sulph.,1-2 gr.	0 28	1 05	0 95
137.	28.	Morphine and Atropine "A"	0 23	0 85	0 75
137.	29.	Morphine and Atropine "B"	0 17	0 60	0 50
	39.	Morphine and Atropine "D" (added).	0 19	0 70	0 60

Morphine sulphate, 1-4 gr.
Atropine sulphate, 1-150 gr.

COMPRESSED TABLETS

PAGE.	No.		Per 1000	Per 500	Per 100
138.	8.	Ammon. brom., 5 grs.	$2 30	$1 20	$0 30
138.	9.	Ammon. brom.,10 grs.	2 65	1 35	0 33
138.	10.	Ammon. chlor., 3 grs.	0 65	0 35	0 15
138.	11.	Ammon. chlor., 5 grs.	0 75	0 40	0 15
138.	12.	Ammon. chlor.,10 grs.	1 00	0 53	0 15
139.	17.	Anti-Pain	2 00	1 05	0 25
139.	18.	Antipyrine, 3 grs.	5 00	2 53	0 55

COMPRESSED TABLETS
(*Continued*)

Page.	No.		Per 1000	Per 500	Per 100
139.	19.	Antipyrine, 5 grs . .	$7 90	$4 00	$0 85
139.	20.	Antipyrine, 10 grs. .	14 75	7 40	1 50
140.	24.	Borax, 5 grs.	0 70	0 38	0 15
140.	25.	Boric acid, 5 grs. . .,	0 85	0 45	0 15
140.	26.	Calomel and Soda . .	1 65	0 85	0 20
140.	27.	Cathartic Compound .	1 80	0 93	0 21
140.	30.	Dover's Powder, 2 grs.	1 60	0 83	0 19
140.	31.	Dover's Powder, 3 grs.	1 95	1 00	0 23
140.	32.	Dover's Powder, 5 grs.	3 00	1 53	0 33
141.	45.	Pepsin, sacch., 5 grs.	1 25	0 65	0 16
141.	50.	Phenacetine, 1 gr . .	6 00	3 03	0 63
141.	51.	Phenacetine, 2 grs. .	10 65	5 35	1 10
141.	52.	Phenacetine, 3 grs. .	16 00	8 05	1 65
141.	58.	Potass. brom., 5 grs. .	1 80	0 93	0 21
141.	59.	Potass. brom., 10 grs.	3 20	1 63	0 35
142.	61.	Pot. chlor. and Borax	1 00	0 53	0 15
142.	62.	Pot. chlor. and Borax Comp.	2 00	1 03	0 24
142.	63.	Potassium iodide, 5 grs.	7 40	3 73	0 78
142.	64.	Pot. permang., 1-2 gr.	0 66	0 36	0 15
142.	65.	Pot. permang., 1 gr. .	0 66	0 36	0 15
142.	66.	Pot. permang., 2 grs.	0 72	0 39	0 15
142.	67.	Pot. permang., 3 grs.	0 80	0 43	0 15
142.	68.	Pot. permang., 5 grs.	1 00	0 53	0 15
142.	77.	Salicin, 2 1-2 grs . .	6 65	3 35	0 70
142.	78.	Salicin, 5 grs. . . .	11 65	5 85	1 20
142.	79.	Salicylic acid, 2 1-2 grs	1 25	0 65	0 16
142.	80.	Salicylic acid, 5 grs. .	1 80	0 93	0 22
143.	82.	Salol, 2 1-2 grs. . . .	2 35	1 20	0 29
143.	83.	Salol, 5 grs.	4 20	2 13	0 47
143.	84.	Soda Mint	0 70	0 38	0 15
143.	86.	Sodium bicarb., 5 grs.	0 65	0 5	0 15
143.	87.	Sodium bicarb., 10 grs.	0 80	0 43	0 15
143.	88.	Sodium brom., 5 grs.	1 90	0 98	0 .23
143.	89.	Sodium brom., 10 grs.	3 40	1 73	0 38
143.	90.	Sodium salicyl., 3 grs.	1 50	0 80	0 20
143.	91.	Sodium salicyl., 5 grs.	2 05	1 05	0 25
143.	92.	Strychnine, 1-120 gr.	0 65	0 35	0 15
143.	94.	Strychnine, 1 60 gr. .	0 65	0 35	0 15

6

SOFT ELASTIC CAPSULES

7

SOFT ELASTIC CAPSULES FILLED WITH POWDERS

			1 doz., 12 in box.	1 doz., 24 in box.	100 in bulk.
PAGE.	No.				
156.	111.	Antipyrine, 3 grs. .	$3 25	$6 20	$1 90
156.	112.	Antipyrine, 5 grs. .	4 00	7 70	2 45
156.	113.	Salol, 2 1-2 grs. . .	1 65	3 00	0 90
156.	114.	Salol, 5 grs. . . .	2 40	4 50	1 40

MEDICINAL ELIXIRS

Elixirs are furnished in 1-pint, 5-pint and gal. bottles.

All listed at $ 9.00 per doz. pints are $4.50 per gal.
"   "   9.75   "   "   "   4.75   "
"   "   10.00   "   "   "   5.00   "
"   "   10.50   "   "   "   5.50   "
"   "   12.00   "   "   "   7.00   "

			Per doz. pints.	Per gallon.
PAGE.	No.			
157.	5.	Ammonium valerianate . .	$10 00	$5 00
163.	52.	Calisaya (Cinchona) . . .	7 50	4 00
163.	53.	Calisaya, ferrated	7 50	4 00
159.	24.	Calisaya and Bismuth . . .	7 50	4 00
160.	28.	Calisaya, Iron and Bismuth	7 50	4 00
160.	29.	Calisaya, Iron, Bismuth and Pepsin	12 00	7 00
160.	30.	Calisaya, Iron, Bismuth, Pepsin and Strychnine .	12 00	7 00
161.	36.	Calisaya, Iron and Strychnine	7 50	4 00
161.	37.	Calisaya, Iron, Strychnine and Pepsin	10 00	5 00
161.	38.	Calisaya, Pepsin and Bismuth	10 00	5 00
161.	39.	Calisaya, Pepsin, Bismuth and Strychnine	10 00	5 00
161.	40.	Calisaya, Pepsin and Strychnine	10 00	5 00
161.	42.	Calisaya and Strychnine . .	7 50	4 00
162.	48.	Celery and Guarana . . .	12 00	7 00
163.	52.	Cinchona, "detannated" (alkaloids)	7 50	4 00

MEDICINAL ELIXIRS

(*Continued*)

PAGE.	No.		Per doz. pints.	Per gallon.
163.	53.	Cinchona, ferrated	$ 7 50	$4 00
163.	57.	Corydalis Comp.	8 00	4 25
164.	66.	Grindelia aromatic	8 00	4 25
164.	69.	Iron phosphate	8 00	4 25
164.	70.	Iron and Quinine, "A" .	9 75	4 75
165.	71.	Iron and Quinine, "B" .	9 75	4 75
165.	72.	Iron and Quinine, "C" .	9 75	4 75
165.	73.	Iron, Quinine and Arsenic.	9 75	4 75
165.	74.	Iron, Quinine and Strychnine, "A".	9 75	4 75
165.	75.	Iron, Quinine and Strychnine, "B"	9 75	4 75
165.	78.	Iron and Strychnine, "A"	9 00	4 50
165.	79.	Iron and Strychnine, "B"	9 00	4 50
166.	80.	Kola Compound.	9 00	4 50
166.	81.	Lactinated Pepsin	9 20	5 00
166.	82.	Lactinated Pepsin and Bismuth	10 00	5 00
166.	83.	Lactinated Pepsin, Bismuth and Strychnine	10 00	5 00
167.	89.	Manaca and Salicylates . .	13 50	7 50
167.	94.	Pepsin and Bismuth . . .	11 00	6 00
169.	107.	Pepsin and Wafer Ash .	12 75	7 25
169.	111.	Potassium bromide. . . .	8 00	4 25
170.	115.	Sodium bromide.	8 00	4 25

MEDICINAL SYRUPS

All listed at $ 9.00 per doz. pints are $4.50 per gal.
  "   "   9.75   "   "   "   4.75   "
  "   "   10.00   "   "   "   5.00   "
  "   "   10.50   "   "   "   5.50   "

PAGE.	No.		Per doz. pints.	Per gallon.
171.	4.	Calcium hypophosphite . .	$ 8 00	$4 25
171.	6.	Calcium lactophosphate . .	11 00	6 00
171.	7.	Calcium and Sodium lactophosphate	12 00	7 00

MEDICINAL SYRUPS

(*Continued*)

PAGE.	No.		Per doz. pints.	Per gallon.
171.	8.	Calc. lactophos. Comp. with Iron.	$12 75	$7 25
172.	11.	Hydriodic acid, U. S. P. .	10 00	5 00
173.	17.	Ipecac, U. S. P.	12 00	7 00
173.	19.	Iron and Manganese iodides	12 00	7 00
173.	20.	Iron, Quinine and Strychnine phosphates	11 00	6 00
173.	26.	Manganese iodide	12 75	7 25
173.	27.	Mitchella Compound . . .	7 20	3 80
174.	32.	Rhubarb	7 20	3 80
174.	35.	Sarsaparilla Comp., U.S.P.	7 20	3 80
175.	39.	Sodium hypophosphite . .	8 00	4 25
175.	40.	Squill, U. S. P.	6 00	3 00
175.	41.	Squill Comp., U. S. P. . .	7 20	3 80
175.	42.	Stillingia Comp.	7 20	3 80
175.	44.	Tolu, U. S. P.	7 20	3 80
175.	45.	Trifolium Comp.	10 25	5 40
176.	47.	Wild Cherry	7 20	3 80
176.	48.	Yerba Santa Aromatic. . .	7 20	3 80

MEDICINAL WINES

PAGE.	No.		Per doz. pints.	Per gallon.
177.	4.	Beef	$ 8 00	$4 25
178.	12.	Coca and Beef	12 00	7 00
178.	13.	Coca, Beef and Iron . . .	12 00	7 00
178.	15.	Colchicum seed	12 00	7 00
178.	17.	Ipecac, U. S. P.	18 00	9 00
178.	19.	Iron, bitter, U. S. P. . . .	8 00	4 25
178.	21.	Iron citrate, U. S. P. . . .	8 00	4 25

OINTMENTS

Ointments are in tin containers, except where otherwise specified.

PAGE.	No.		Per ℔.
192.	22.	Mercurial, 1-2 mercury	$1 25
192.	23.	Mercurial, 1-3 mercury	1 00
192.	42.	Zinc oxide, U. S. P.	1 00

PART II.

BIOLOGIC PRODUCTS

Page 194. Stearns' Antistreptococcic Serum, per package, $3.00. Each package contains 20 Cc. in Syro-Bulbs. (*Discount 30%*)

Page 195. Stearns' Diphtheritic Antitoxin, Regular or Standard strength (250 to 500 units per Cc.), is now furnished only in the Syro-Bulb.

SPECIALTIES

Page 199. Liquid Hæmoferrum. Each teaspoonful contains four grains Hæmoferrum.

Page 200. Vibutero, per doz., $6.00 *Net*.

PATENTED

ERRATA

PILLS

No. 53 is sugar-coated only.

Nos. 318, 325, 326 are gelatin-coated only.

Nos. 11, 12, 13, 15, 16, 19, 20, 24, 229, 230, 232 are furnished in both sugar and gelatin coatings.

CAPSOIDS

Santal Compound contains: Santal oil, 2 m.; Copaiba oil, 2 m.; Cubeb oil, 1 m.

THE

PHARMACEUTICAL PRODUCTS

OF

FREDERICK STEARNS & CO.,

MANUFACTURING PHARMACISTS,

DETROIT, MICH., U. S. A.

CANADIAN LABORATORY:

WINDSOR, ONTARIO.

BRANCHES AT

LONDON, ENG.,   NEW YORK CITY,
     25 Lime St.   32 Platt St.

FOREIGN AGENCIES:

AUSTRALIA.
Gibbs, Bright & Co. . . Melbourne, Sydney and Brisbane.
W. R. Cave & Co. . . . . Adelaide.
NEW ZEALAND.
New Zealand Drug Co., Ltd.   Sharland & Co., Ltd.
TASMANIA.
L. Fairthorne & Son . . . . Launceston.
HAWAIIAN ISLANDS.
Hollister Drug Co., Ltd. . . . . Honolulu.
INDIA.
J. A. Kirkbride & Co. . . . . Bombay.
Buhrer, Boeckel & Co. . . . . Calcutta.
Leighton & Co. . . . . Madras.
Darley, Butler & Co. . . . . Colombo.
A. Scott & Co. . . . . Rangoon.
EGYPT.
Max Fischer . . . . Alexandria and Cairo.

CABLE ADDRESS—"STEARNS DETROIT."

NOTES OF INFORMATION.

We shall be pleased to supply special quotations on such products as may be desired in bulk, or in large quantities.

Funds in settlement must be at par in Detroit, New York, Chicago or Windsor.

All claims will have prompt attention if made immediately on receipt of goods; but if such are delayed until an account is due, or until too late to trace, they cannot be entertained.

Breakage, leakage or other damage by careless handling, and all risks by fire, water, collision, explosion, robbery, or otherwise, while goods are in transit, are at risk of buyer.

We cannot supply loose labels for preparations of our manufacture, as we are held responsible for the identity and quality of our products, and are forced to adopt this arbitrary rule.

FREDERICK STEARNS & CO.,

WINDSOR, ONT. ⎫
LONDON, ENG. ⎬ DETROIT, MICH.; U. S. A.
NEW YORK CITY. ⎭

ASSAYED FLUID EXTRACTS.

The advantages of ASSAYED FLUID EXTRACTS over those which are not standardized are so apparent that special mention seems unnecessary. Our line of these preparations, listed in the following pages, we believe to be the most complete offered by any American house, while the methods employed in their manufacture, from the selection of the crude drugs to the final assay of the finished products, are such as to justify us in the claim that our ASSAYED FLUID EXTRACTS are unsurpassed for therapeutic activity, uniformity of strength and pharmaceutical elegance. For further information see note 49.

This list will be found to contain several novel features, chief among which is the statement of medical properties and dose of each fluid extract. The botanical titles are given in italics in all cases, but as there has seemed to be no good reason for appending the authors' names, these have been omitted. The descriptions of medical properties, dose and botanical origin of the drug have been compiled from the most authentic sources, and this list may be consulted for information on any of these points with the certainty that it is thoroughly reliable.

As a matter of convenience, a number of tinctures, half-strength fluids and compound fluids are listed along with the fluid extracts; in every case, however, these are appropriately designated. The formula of each compound fluid is given in statements of metric and ordinary measures.

The list price is ten cents per pound *less* when five-pound bottles are ordered; ten cents per pound *more* when half-pound bottles are ordered; and twenty cents per pound *more* when quarter-pound bottles are ordered. Unless otherwise directed, we will supply pound bottles on all orders. The word pound is here used in the ordinary commercial acceptation of the term, meaning one (wine) pint of sixteen fluidounces.

An exhaustive list of synonyms, both botanical and popular, will be found in the last pages of this Catalogue.

Per lb.

1. Abscess-root, *Polemonium reptans* . . . $1 50
   Astringent, diaphoretic, expectorant.
   Dose, 30 to 60 minims.

2. Absinthium, *Artemisia Absinthium* . . . 1 00
   Stomachic, anthelmintic, antiperiodic.
   Dose, 10 to 30 minims.

3. Aconite leaves, assayed, *Aconitum Napellus* 1 10
   Standard; 1:100 dilution, physiological test.
   Anodyne, sedative, antipyretic.
   See note 49. Dose, 2 to 4 minims.

4. Aconite root, assayed, *Aconitum Napellus* 1 20
   Standard: 1:700 dilution; physiological test.
   Anodyne, sedative, antipyretic.
   See note 49. Dose, ½ to 1 minim.

5. *Adhatoda vasica*, Malabar nut 3 00
   Germicide, expectorant, antispasmodic.
   See note 3. Dose, 15 to 60 minims.

6. *Adonis vernalis*, False Hellebore 3 00
   Cardiac stimulant, diuretic. See note 4.
   Dose, 1 to 2 minims.

7. Adrue, *Cyperus articulatus* 3 50
   Tonic, antemetic.
   Dose, 20 to 30 minims.

8. Agaric, White, *Polyporus officinalis* . . 3 00
   Antisudorific, cathartic.
   Dose, 5 to 30 minims.

9. Agrimony, *Agrimonia Eupatoria* 1 00
   Astringent, tonic.
   Dose, 30 to 60 minims.

10. *Ailanthus glandulosa*, Tree of Heaven . 1 90
    Depressant, antispasmodic.
    Dose, 15 to 30 minims.

    Aletris farinosa (474). See Unicorn-root 1 80

11. Alkanet, *Alkanna tinctoria* 1 50
    Coloring agent.

    Allspice (359). See Pimenta 1 25

12. Aloes, Socotrine, *Aloe Perryi* 1 65
    Cathartic, emmenagogue. See note 51.
    Dose, 2 to 5 minims.

13. Aloes, for U. S. P. Tincture 1 65
    100 Cc. represent—   1 fl. oz. represents—
      30 Gm. Purified Aloes . . . . 137 grs.
      60 Gm. Glycyrrhiza . . . . 274 grs.
    Three times as strong as the Tincture.
    See note 50.

Per lb.

14. Aloes and Myrrh, for U. S. P. Tincture . $1 75
    100 Cc. represent—   1 fl. oz. represents—
        30 Gm. Purified Aloes . . . . 137 grs.
        30 Gm. Myrrh . . . . 137 grs.
        30 Gm. Glycyrrhiza . . . . . 137 grs.
    Three times as strong as the Tincture.
    See note 50. . .

15. Aloes Comp., for B. P. Decoction . . .   1 75
    100 Cc. represent—   1 imp. pint represents—
        5. Gm. Ext. Socotrine Aloes . . 1 oz.
        2.5 Gm. Myrrh . . . . ½ oz.
        2.5 Gm. Saffron . . . . ½ oz.
        2.5 Gm. Potassium carbonate. . ½ oz.
        20. Gm. Ext. Licorice . . . . 4 ozs.
        37.5 Cc. Comp. Liq. Cardamom.7½ ozs.
    Five times as strong as the Decoction.
    See note 50.

16. Aloes, for B. P. Tincture 1 65
    100 Cc. represent—   1 imp. pint represents—
        12.5 Gm. Socotrine Aloes . . . 2½ ozs.
        37.5 Gm. Ext. Licorice . . . .7½ ozs.
    Five times as strong as the Tincture.
    See note 50.

17. Alterative Comp. (Bamboo-brier Comp.)   1 25
    100 Cc. represent—   1 fl. oz. represents—
        26.33 Gm. Smilax Sarsaparilla. 120 grs.
        26.33 Gm. Stillingia . . . . . . 120 grs.
        26.33 Gm. Lappa . . . . 120 grs.
        13.17 Gm. Poke root . . . . . 60 grs.
        13.17 Gm. Xanthoxylum . . . 60 grs.
    Alterative, tonic, antisyphilitic.
    See note 5. Dose, 60 to 120 minims.

    Althaea (314). See Marshmallow . . .   1 00

18. American Centaury, *Sabbatia angularis* . 1 00
    Tonic, antiperiodic. Dose, 60 minims.

19. American Columbo, *Frasera Walteri* . . 1 25
    Bitter tonic. Dose, 60 minims.
    American Hellebore, assayed(478). See
        Veratrum viride 1 50
    Standard: 1.% total alkaloids, by weight.

20. American Ivy, *Ampelopsis quinquefolia* . 1 20
    Tonic, expectorant, astringent.
    Dose, 30 to 60 minims.

21. American Saffron, *Carthamus tinctorius*. 2 40
    Laxative, diaphoretic.
    Dose, 8 to 15 minims.

Per lb.

22. American Sarsaparilla, *Aralia nudicaulis.* $1 05
     Alterative,   Dose, 30 to 60 minims.
     American Wormseed (513).   See Worm-
       seed, American 1 00

23. Angelica root, *Angelica Archangelica* . . 1 20
     Tonic, carminative, stomachic.
     Dose, 30 to 60 minims.

24. Angelica seed, *Angelica atropurpurea* . 1 50
     Tonic, carminative, stimulant.
     Dose, 30 to 60 minims.

25. Angustura, *Galipea Cuspária* 1 50
     Aromatic bitter.   Dose, 8 to 30 minims.

26. Anise, *Pimpinella Anisum* 1 50
     Stomachic, carminative.
     Dose, 10 to 30 minims.

     Anthemis (139). See Chamomile, Roman 1 20
     Apocynum (108). See Canadian Hemp . 1 20

27. Apple-tree bark, *Pyrus malus* 1 20
     Tonic, febrifuge.   Dose, 15 to 60 minims.

28. Arbor vitæ, *Thuja occidentalis* 1 20
     Uterine stimulant, diuretic.
     Dose, 15 to 60 minims.

29. Arbor vitæ, aqueous 1 20
30. Areca nut, *Areca Catechu* 2 25
     Astringent, tonic, vermifuge.
     Dose, 60 to 240 minims.

31. Arnica flowers, *Arnica montana* 1 20
     Stimulant, diuretic, vulnerary.
     See note 7.   Dose, 5 to 10 minims.

32. Arnica root, *Arnica montana* 1 20
     Stimulant, diuretic, vulnerary,
     See note 7.   Dose, 5 to 10 minims.

33. Aromatic 2 50
     100 Cc. represent—   1 fl. oz. represents—
       .35 Gm. Ceylon Cinnamon . . . 160 grs.
       35 Gm. Ginger . . . . 160 grs.
       15 Gm. Cardamom . . . . . . 68 grs.
       15 Gm. Nutmeg . . . . 68 grs.
     Carminative, adjuvant.   See note 50.
     Dose, 10 to 20 minims.

34. Asafetida, *Ferula foetida* 1 50
     Antispasmodic, expectorant, laxative.
     See note 51.   Dose, 5 to 15 minims.

Per lb.

Asclepias (367). See Pleurisy-root· . . 1 20

35. Asparagus, *Asparagus officinalis* 1 50
    Diuretic, laxative, cardiac sedative,
    Dose, 30 to 60 minims.

    Aspidium, assayed··(308).· ; See ·Male·
    .fern .. .. ·\`\`\`\`\`'... . 1 25
        Standard: 10.% oleoresin. . .
    Aspidosperma, assayed·(379)·.. ·See Que-
    bracho ·.". 2 50
        Standard: 1.% alkaloids, by weight.· ·

36. Bael fruit, ·*Aegle Marmelos* 2 25
    Astringent. Dose, 15 to 60 minims.

37. Balm, *Melissa officinalis*,.. . . .. . 1 20
    Carminative, diaphoretic, stimulant.
    Dose, 15 to 60 minims.·· . ... · · ·

    Balm of·Gilead 'buds (40)·.· See Balsam
    ·Poplar buds 1 50

38. Balmony, *Chelone glabra*. 1 00
    ·· Tonic, aperient, anthelmintic.
    Dose, 30 to 60 minims. ··

39. Balsam Fir bark,·*Abies balsamea* . ., . 1 20
    Stimulant diuretic. Dose, 30 to 60 minims.

40. Balsam Poplar buds, ·*Populus balsami-
    fera*, var. *candicans* 1 50
    Terebinthinate, diuretic.
    Dose, 60 to·120 minims..

41. Bamboo-brier root, *Smilax Sarsaparilla* . 1 50
    Alterative. , See note 8.
    Dose, 60 to 120 minims.

    Bamboo-brier·Comp. (17). See Altera-
    tive Comp.· . . . . . ·.· . . . ·. . 1 25

42. Barberry, *Berberis vulgaris* 1 20
    Alterative, tonic. Dose, 3 to 10 minims.

43. Bayberry, *Myrica cerifera* 1 00
    Astringent, tonic. Dose, 15 to 30 minims.

44. Bay Laurel, Conc.,*Myrcia acris* 3 00
    For Bay Rum. See note 9.

45. Bearsfoot, *Polymnia Uvedalia* 1 75
    External; discutient, anodyne.

46. Beech-drop, *Epiphegus Virginiana* . . . 1 50
    Astringent. Dose, 30 to 60 minims.

Per lb.

47. Belladonna leaves, assayed, *Atropa Bella-*
donna. $1 25
Standard : .4% alkaloids, by titration.
Narcotic, diuretic, sedative.
See note 49.   Dose, 1 to 3 minims.

48. Belladonna root, assayed, *Atropa Bella-*
donna. 1 50
Standard : .5% alkaloids, by titration.
Narcotic, diuretic, sedative.
See note 49.   Dose, 1 to 2 minims.

49. Benne leaves, *Sesamum Indicum* . . . 1 50
Astringent, demulcent.   Dose 1 to 10 min.

50. Benzoin, for U: S: P. Tincture, *Styrax*
Benzoin. 2 00
Stimulant, expectorant.   See note 51.
Dose, 30 to 60 minims.

51. Benzoin Comp., for U. S. P. Tincture . 2 00
100 Cc. represent—   1 fl. oz. represents—
32.   Gm. Benzoin. . . . . . . 146 grs.
5.33 Gm. Purified Aloes . . . 24 grs.
21.33 Gm. Storax . . . . . . 97 grs.
10.67 Gm. Balsam Tolu . . . . 49 grs.
Two and two-thirds times as strong as the
Tincture.   See note 50.

52. Benzoin Comp., for B. P. Tincture. . . 2 00
100 Cc. represent— 1 imp'l pt. represents—
30.   Gm. Benzoin . . . . . 6 ozs.
22.5 Gm. Prepared Storax . 4½ ozs.
7.5 Gm. Balsam Tolu . . . 1½ ozs.
5.48 Gm. Socotrine Aloes . . 480 grs.
Three times as strong as the Tincture.
See note 50.

53. *Berberis Aquifolium*, assayed, Oregon
grape. 2 40
Standard : 3.25% berberine, by titration.
Tonic, alterative.
See notes 11, 49.   Dose, 8 to 30 minims.

Betel nut (30).   See Areca nut 2 25

54. Beth-root, *Trillium erectum* 1 05
Emmenagogue, emetic.   Dose, 15 to 60 m.

55. Bistort, *Polygonum Bistorta* 1 50
Tonic, astringent.   Dose, 8 to 30 minims.

56. Bitter Bugleweed, *Lycopus Europaeus* . 2 00
Pectoral.   Dose, 10 to 30 minims.

Bitter Orange peel (339).   See Orange
peel, Bitter. 1 40

Per lb.

57. Bitter-root, *Apocynum androsaemifolium* $1 20
Cathartic, emetic, diaphoretic.
Dose, 5 to 20 minims.

58. Bittersweet, *Solanum Dulcamara* . . . 1 20
Alterative, resolvent.
Dose, 60 to 120 minims.

59. Black Alder, *Ilex verticillata* 1 00
Tonic, antiperiodic, astringent.
Dose, 15 to 60 minims.

60. Black Ash bark, *Fraxinus sambucifolia* . 1 20
Tonic, astringent. Dose, 60 to 120 minims.

61. Blackberry-root bark, *Rubus villosus* . . 1 00
Tonic, astringent. Dose, 15 to 30 minims.

62. Blackberry Aromatic, for Elixir 1 20
100 Cc. represent— 1 fl. oz. represents—
71.93 Gm. Blackberry-root . . 328 grs.
6.58 Gm. Cardamom . . . . . 30 grs.
6.58 Gm. Cinnamon . . . . 30 grs.
3.29 Gm. Ginger . . . . . . 15 grs.
3.29 Gm. Caraway . . . . . 15 grs.
Four times as strong as the Elixir.
See notes 12, 50.

63. Black Cohosh, *Cimicifuga racemosa* . . 1 20
Antispasmodic, diaphoretic, expectorant.
Dose, 15 to 30 minims.

64. Black Cohosh Comp., for Elixir 1 20
100 Cc. represent— 1 fl. oz. represents—
32.9 Gm. Cimicifuga . . . . . 150 grs.
32.9 Gm. Wild Cherry . . . . 150 grs.
16.45 Gm. Senega . . . . . 75 grs.
8.22 Gm. Ipecac . . . . . . 37 grs.
16.45 Gm. Licorice . . . . . 75 grs.
Eight times as strong as the Elixir.
See notes 13, 50.

65. Black Haw, *Viburnium prunifol.* . . 1 20
Uterine tonic and sedative.
Dose, 30 to 60 minims.

66. Black Hellebore, *Helleborus niger.* . . 1 20
Hydragogue cathartic.
Dose, 5 to 15 minims.

67. Black Pepper, *Piper nigrum* 2 00
Carminative, condiment.
Dose, 5 to 20 minims.

68. Black Walnut leaves, *Juglans nigra* . . 2 00
Tonic, laxative. Dose, 60 to 120 minims.

Per lb.

69. Black Willow bark, *Salix nigra* $1 50
       Tonic, astringent, febrifuge. See note 14.
       Dose, 15 to 60 minims.

70. Black Willow buds, *Salix nigra* 1 50
       Tonic, astringent, febrifuge. See note 14.
       Dose, 15 to 60 minims.

71. Bladder Wrack, *Fucus vesiculosus* . . . 1 35
       For reducing obesity.
       Dose, 30 to 120 minims.

72. Blessed Thistle, *Cnicus benedictus* . . . 1 20
       Tonic, diaphoretic, emetic.
       Dose, 15 to 60 minims.

73. Blood-flower, *Asclepias curassavica* . . . 2 50
       Emetic, cathartic, vermifuge. See note 15.
       Dose, 60 to 120 minims.

74. Blood-root, assayed, *Sanguinaria Cana-
       densis* 1 05
       Standard; 3.5% total alkaloids, by titration.
       Expectorant, stimulant, emetic, narcotic.
       See note 49. Dose, 2 to 30 minims.

75. Blue Cohosh, *Caulophyllum thalictroides* 1 00
       Sedative, diuretic, emmenagogue.
       Dose, 5 to 30 minims.

76. Blue Cohosh Comp., for Eclec. Disp.
       Tincture. 1 20
       100 Cc. represent— 1 fl. oz. represents—
            46.70 Gm. Blue Cohosh . . . . 213 grs.
            23.35 Gm. Ergot. . . . . . . 107 grs.
            23.35 Gm. Water Pepper . . . 107 grs.
            11.11 Cc. Oil Savine . . . . . 53 m.
       Five and one-third times as strong as the
            Tincture. See note 50.

77. Blue Flag, *Iris versicolor* 1 05
       Cholagogue cathartic.
       Dose, 10 to 30 minims.

78. Boldo, *Peumus Boldus* 3 00
       Stimulant, tonic, diuretic. See note 16.
       Dose, 15 to 60 minims.

79. Boneset, *Eupatorium perfoliatum* 1 00
       Diaphoretic, laxative, emetic.
       Dose, 20 to 60 minims.

80. Borage, *Borago officinalis* 1 00
       Diuretic, demulcent. Dose, 60 minims.

81. Broom, *Cytisus Scoparius* 1 05
       Diuretic. Dose, 15 to 60 minims.

Per lb.

82. Broom Corn seed, *Sorghum saccharatum*. $1.20
Diuretic, lithontriptic. See note 17.
Dose, 30 to 60 minims.

Bryonia (492). See White Bryony. 1 50

83. Buchu, *Barosma betulina* 1 50
Diuretic. See note 18.
Dose, 15 to 60 minims.

84. Buchu Comp., Stearns', for Elixir, . 1 50
100 Cc. represent— 1 fl. oz. represents— 1 20
26.33 Gm. Buchu. . . . . 120 grs.
26.33 Gm. Cubeb. . . . . 120 grs.
26.33 Gm. Juniper berries . . . 120 grs.
26.33 Gm. Uva Ursi . . . . . . 120 grs.
Four times as strong as the Elixir.
See notes 19, 50.

85. Buchu Comp., for N. F. Elixir . . . 1 50
100 Cc. represent— 1 fl. oz. represents—
62.5 Gm. Buchu . . . . 285 grs.
12.5 Gm. Cubeb. . . . . 57 grs.
12.5 Gm. Juniper berries. . . . 57 grs.
12.5 Gm. Uva Ursi . . . . . 57 grs.
Four times as strong as the Elixir.
See notes 19, 50.

86. Buchu, Juniper and Potassium acetate, for
Elixir 1 50
100 Cc. represent— 1 fl. oz. represents—
79.00 Gm. Buchu . . . . 360 grs.
19.75 Gm. Juniper berries . . . 90 grs.
13.17 Gm. Potassium acetate . 60 grs.
Four times as strong as the Elixir.
See notes 19, 50.

87. Buchu and Pareira Brava, for Elixir . 1 50
100 Cc. represent— 1 fl. oz. represents—
52.67 Gm. Buchu . . . . 240 grs.
26.33 Gm. Pareira Brava . . . 120 grs.
26 33 Gm. Stone-root . . . . . 120 grs.
Four times as strong as the Elixir.
See notes 19, 50.

88. Buchu and Pareira Comp., for Elixir . 1 50
100 Cc. represent— 1 fl. oz. represents—
13.17 Gm. Buchu . . . . 60 grs.
13.17 Gm. Dandelion . . . . . . 60 grs.
8.78 Gm. Juniper berries . . . 40 grs.
8.78 Gm. Stone-root . . . . . 40 grs.
8.78 Gm. Pareira Brava . . . 40 grs.
6.58 Gm. Potassium acetate . . 30 grs.
Twice as strong as the Elixir.
See notes 19, 50.

Per lb.

89. Buckbean, *Menyanthes trifoliata* $1 20
Bitter tonic, cathartic.
Dose, 15 to 45 minims.

90. Buckeye bark, *Aesculus glabra* 1 20
Relieves portal congestion.
Dose, 3 to 5 minims.

91. Buckhorn brake, *Osmunda regalis* . . . 1 20
Demulcent, tonic, styptic.
Dose, 60 to 120 minims.

92. Buckthorn bark, *Rhamnus Frangula* . . 1 00
Tonic, purgative.
Dose, 15 to 30 minims.

93. Buckthorn berries, *Rhamnus cathartica* . 1 05
Drastic purgative.
Dose, 15 to 30 minims.

94. Bugleweed, *Lycopus Virginicus* 1 00
Astringent, sedative.
Dose, 10 to 30 minims.

Bugleweed, Bitter (56). See Bitter
Bugleweed 2 00

95. Burdock root, *Arctium Lappa* 1 00
Diuretic, diaphoretic, alterative.
Dose, 30 to 120 minims.

96. Burdock seed, *Arctium Lappa* 1 20
Tonic, antiscorbutic.
Dose, 10 to 30 minims.

97. Butternut, *Juglans cinerea* 1 00
Tonic, cathartic. Dose, 60 to 120 minims.

98. Button Bush, *Cephalanthus occidentalis* . 1 20
Febrifuge, laxative, diuretic.
Dose, 30 to 60 minims.

99. Button Snakeroot, *Lacinaria spicata* . . 1 00
Diuretic. Dose, 30 to 120 minims.

100. Cabbage-tree bark, *Andira inermis* . . . 4 50
Cathartic, vermifuge. Dose, 15 to 30 minims.

Cactus grandiflorus (135). See *Cereus
grandiflorus* 4 00

101. Calabar Bean, assayed, *Physostigma ven-
enosum* 3 60
Standard: .25% physostigmine, by weight.
Spinal sedative, myotic. See note 49.
Dose, 1 to 3 minims.

102. Calamus, *Acorus Calamus* 1 25
Carminative, tonic.
Dose, 15 to 60 minims.

Per lb.

103. *Calendula officinalis,* Marigold . . ; ;.. .. $2 20
Stimulant, diaphoretic, vulnerary.
.. Dose, 15 to 60 minims. . .'

104. California Fever Bush, *Garrya Fremontii.* 3 50
Tonic, antiperiodic. . Dose, 15 to 30 minims.

105. California. Laurel, *Umbellularia Cali-
fornica* . . . .:,. . . . 5 00
Aromatic, anodyne. See note 21.
Dose, 10 to 30 minims.
Calumba (168). See Columbo.. . . . 1 20

106. Camphor Comp., for B. P. Tincture . . 1 50
100 Cc. represent— 1 imp. pt. represents—
3.65 Gm. Opium .. . . 320 grs.
3.65 Gm. Benzoic acid . 320 grs. . .,
2.74 Gm. Camphor . . 240 grs.
2.50 Cc. Oil Anise . . 4 fluidrachms.
Eight times as strong as the Tincture.
See notes 50, 82.

107. Canada Snakeroot, *Asarum Canadense .* 1 05
Aromatic stimulant, tonic. ,,.
Dose, 20 to 30 minims. ,

108. Canadian Hemp, *Apocynum cannabinum* 1 20
Diuretic, emetic, diaphoretic, cathartic.
See note 63. Dose, 5 to 20 minims.

109. Canella, *Canella alba* . . . . . .:. . . 1 00
Aromatic, stimulant.
Dose, 10 to 40 minims. ,.

110. Cannabis Indica, *Cannabis sativa* 1 50
Anodyne, narcotic., See note 63.
Dose, 1 to 3 minims. .

111. Cantharides, *Cantharis vesicatoria* . . . 5 50
Vesicant. External. , ,. ' 'r .'

112. Capsicum, *Capsicum fastigiatum .'.* . . . 1 50
Stimulant, condiment, rubefacient.
Dose, ½ to 3 minims. 1

113. Capsicum and Myrrh, for N. F. Tincture . 2 00
100 C.c. represent— 1 fl. oz. represents—
12.8 Gm. Capsicum . . . 58 grs.
50. Gm. Myrrh . . .:.. . . 228 grs.
Four times as strong as the Tincture.
See note 50.

114. Caraway, *Carum carvi* . ...,,.. . . 1 35
Carminative, stimulant. Dose, 8 to 30 minims.

115. Cardamom, *Elettaria repens* . . :.. ... 3 00
Aromatic, carminative, stomachic. , m.,
Dose, 5 to 15 minims. .

Per lb.

116. Cardamom Comp., for U. S. P. Tincture. $2 00
    100 Cc. represent— 1 fl. oz. represents—
      16. Gm. Cardamom . . . . . . 73 grs.
      16. Gm. Cassia Cinnamon . . . . 73 grs.
      8. Gm. Caraway . . . . . . 36 grs.
      4. Gm. Cochineal . . . . . . 18 grs.
    Eight times as strong as the Tincture.
    See note. 50.

117. Cardamom Comp., for B. P. Tincture . . 2 00
    100 Cc. represent—1 imp. pt. represents—
      5. Gm. Cardamom seeds . . . 1 oz.
      5. Gm. Caraway fruit . . . . 1 oz.
      40. Gm. Raisins . . . . . . 8 ozs.
      10. Gm. Cinnamon bark . . . ½ ozs.
      2.5 Gm. Cochineal . . . . . 220 grs.
    Four times as strong as the Tincture.
    See note 50.

    Carduus benedictus (72). See Blessed
    Thistle 1 20

118. Caroba, *Jacaranda procera* 3 50
    Alterative, antisyphilitic. See note 22.
    Dose, 15 to 60 minims.

119. Carpenter's-square, *Scrophularia nodosa*. 1 80
    Diuretic, depurative, anodyne.
    Dose, 30 to 60 minims.

120. Cascara Amarga, *Picramnia*——. 3 00
    Antisyphilitic. See note 23.
    Dose, 30 to 60 minims.

121. Cascara Sagrada, *Rhamnus Purshiana* . 1 50
    Tonic laxative. See note 24.
    Dose, 15 to 60 minims.

    Cascara Aromatic. See "Kasagra,"
    Specialty List, Part II.

122. Cascara Sagrada Comp., for N. F. Elixir . 1 80
    100 Cc. represent— 1 fl. oz. represents—
      40. Gm. Cascara Sagrada . . . 182 grs.
      24. Gm. Senna . . . . 109 grs.
      20.8 Gm. Juglans . . . . . . 95 grs.
    Three and one-fifth times as strong as the
    Elixir. See note 50.

123. Cascara Sagrada, Improved 1 50
    Tonic laxative. See note 25.
    Dose, 15 to 60 minims.

124. Cascarilla, *Croton Eluteria* 1 00
    Stimulant, tonic, febrifuge.
    Dose, 15 to 30 minims.

Per lb.

125. Cassia-Cinnamon, *Cinnamomum Cassia* . $1 80
    Carminative, stimulant.
    Dose, 10 to 30 minims.

126. Castor bean, *Ricinus communis* 1 80
    Cathartic. Dose, 15 to 30 minims.

127. Castor leaves, *Ricinus communis* 1 80
    Cathartic, galactagogue.
    Dose, 15 to 60 minims.

128. Catechu, *Acacia Catechu* 1 20
    Astringent. See note 51.
    Dose, 8 to 15 minims.

129. Catechu Comp., for U. S. P. Tincture . . 1 20
    100 Cc. represent— 1 fl. oz. represents—
      53.33 Gm. Catechu . . . . 243 grs.
      26.67 Gm. Cassia Cinnamon . . 122 grs.
    Five and one-third times as strong as the
      Tincture. See note 50.

130. Catechu, for B. P. Tincture 1 20
    100 Cc. represent— 1 imp. pint represents—
      50. Gm. Catechu . . . . 10 ozs.
      20. Gm. Cinnamon bark . . . . 4 ozs.
    Four times as strong as the Tincture.
    See note 50.

131. Catnep, *Nepeta Cataria* 1 00
    Carminative, tonic, diaphoretic.
    Dose, 15 to 60 minims.

    Caulophyllum (75). See Blue Cohosh . 1 00
    Caulophyllum Comp. (76). See Blue
      Cohosh Comp. 1 20

132. Cedron, *Simaba Cedron* 5 00
    Cerebral sedative, antispasmodic, antiperi-
      odic. See note 26.
    Dose, 1 to 8 minims.

133. Celery seed, *Apium graveolens* 1 80
    Diuretic, sedative, carminative.
    Dose, 5 to 15 minims.

134. Celery Comp., for N. F. Elixir 2 00
    100 Cc. represent— 1 fl. oz. represents—
      24.8 Gm. Celery . . . . 113 grs.
      24.8 Gm. Coca . . . . 113 grs.
      24.8 Gm. Kola . . . . 113 grs.
      24.8 Gm. Viburnum prunifolium 113 grs.
    Four times as strong as the Elixir.
    See note 50.

135. *Cereus grandiflorus*, Cactus grandiflorus . 4 00
    Cardiac tonic. See note 27.
    Dose, 2 to 5 minims.

Per lb.

136. Cevadilla seed, *Asagraea officinalis* . . $1 80
External; parasiticide.

137. Ceylon Cinnamon, *Cinnamomum Zeylani-
cum* 1 80
Carminative, stimulant.
Dose, 10 to 20 minims.

138. Chamomile, German, *Matricaria Chamo-
milla* 1 20
Tonic, stimulant, carminative.
Dose, 15 to 60 minims.

139. Chamomile, Roman, *Anthemis nobilis* . 1 20
Tonic, stimulant, carminative.
Dose, 15 to 60 minims.

140. Chapparro Amargoso, *Castella Nicholsoni*. 2 50
Antiseptic, astringent. See note 28.
Dose, 30 to 60 minims.

141. Chekan, *Eugenia Chekan* 4 00
Tonic, expectorant. See note 29.
Dose, 30 to 60 minims.

Chelidonium (219). See Garden Celandine 1 00

Chenopodium (513). See Wormseed,
American 1 00

Cherry bark, Wild (502). See Wild
Cherry 1 05

142. Chestnut leaves, *Castanea dentata* . . . 1 00
Tonic, astringent. Dose, 30 to 120 minims.

143. Chewstick, *Gouania Domingensis* . . . 2 00
Tonic, laxative. Dose, 60 to 120 minims.

144. Chicory, *Cichorium Intybus* 1 05
Tonic, aperient. Dose, 15 to 60 minims.

Chimaphila (364). See Pipsissewa . . 1 00

145. Chirata, *Swertia Chirata* 1 80
Bitter tonic. Dose, 30 to 60 minims.

Cimicifuga (63). See Black Cohosh . . 1 20

146. Cinchona, assayed, *Cinchona Calisaya* . 1 50
Standard: 5.% total alkaloids, 2.5% quinine,
by weight.
Tonic, antiperiodic, febrifuge, mild astringent.
See notes 30, 49. Dose, 15 to 60 minims.

Per lb.

147. Cinchona Aromatic, for Aromatic Wines
and Elixirs $1 80
   100 Cc. represent— 1 fl. oz. represents—
     46.08 Gm. Cinchona . . . . . .210 grs.
     23.04 Gm. Orange peel . . . . 105 grs.
     11.52 Gm. Cinnamon . . . . . 52 grs.
     11.52 Gm. Coriander . . . . . 52 grs.
     3.29 Gm. Anise . . . . 15 grs.
     3.29 Gm. Caraway . . . . . . 15 grs.
     3.29 Gm. Cardamom . . . . . 15 grs.
     3.29 Gm. Cochineal . . . . . 15 grs.
   See notes 30, 50.

148. Cinchona, detannated 1 70
   See note 30.

149. Cinchona, pale, assayed, *Cinchona offic-
inalis* 1 25
   Standard: 3.% total alkaloids, by weight.
   Tonic, antiperiodic, febrifuge, mild astringent.
   See notes 30, 49. Dose, 15 to 60 minims.

150. Cinchona, red, assayed, *Cinchona suc-
cirubra* 1 50
   Standard: 5.% total alkaloids, by weight.
   Tonic, antiperiodic, febrifuge, mild astringent.
   See notes 30, 49. Dose, 15 to 60 minims.

151. Cinchona Comp., for U. S. P. Tincture . 1 50
   100 Cc. represent— 1 fl. oz. represents—
     53.33 Gm. Red Cinchona . . . 243 grs.
     42.67 Gm. Bitter Orange peel . 195 grs.
     10.66 Gm. Serpentaria . . . 49 grs.
   Five and one-third times as strong as the
   Tincture. See notes 30, 50.

152. Cinchona Comp., for B. P. Tincture . . 1 50
   100 Cc. represent— 1 imp. pt. represents—
     40. Gm. Red Cinchona bark . . 8 ozs.
     20. Gm. Bitter Orange peel . . 4 ozs.
     10. Gm. Serpentary rhizome . 2 ozs.
     2.5 Gm. Saffron . . . . . . 220 grs.
     1.25 Gm. Cochineal . . . . . 112 grs.
   Four times as strong as the Tincture.
   See notes 30, 50.

   Cinnamon, Ceylon (137). See Ceylon
   Cinnamon 1 80

153. Cleavers, *Galium Aparine* 1 00
   Aperient, diuretic, alterative.
   Dose, 30 to 60 minims.

	Per lb.

154. Clotbur, *Xanthium strumarium*. . . . $2 00
　　　Hemostatic, styptic.
　　　Dose, 60 to 120 minims.

155. Clover blossoms, Red, *Trifolium pratense* 1 20
　　　Alterative, deobstruent.
　　　Dose, 30 to 120 minims.

156. Cloves, *Eugenia aromatica* 2 10
　　　Aromatic, stomachic. Dose, 4 to 10 minims.

157. Coca, assayed, *Erythroxylon Coca* . . . 2 00
　　　Standard: .5% cocaine, by weight.
　　　Nerve stimulant, anodyne.
　　　See notes 31, 49. Dose, 15 to 60 minims.

158. Coca, soluble 3 00
　　　See notes 31, 50.

159. Cocculus Indicus, *Anamirta paniculata* . 1 20
　　　Externally, parasiticide; internally, powerful
　　　　　nerve stimulant. Dose, ¼ to 1 minim.

160. Cochineal, *Coccus cacti* 2 05
　　　Coloring agent.

161. Cocillaña, *Sycocarpus Rusbyi* 5 00
　　　Expectorant, diaphoretic, emetic.
　　　See note 32. Dose, 10 to 30 minims.

162. Coffee, green, *Coffea Arabica* 1 50
　　　Stimulant, tonic. Dose, 30 to 60 minims.

163. Coffee, roasted, for Syrup 1 50
　　　See note 50.

164. Colchicum root, assayed, *Colchicum au-
　　　tumnale* 1 20
　　　Standard: .5% colchicine, by weight.
　　　Antipodagric, cathartic, emetic.
　　　See note 49. Dose, 1 to 5 minims.

165. Colchicum seed, assayed, *Colchicum au-
　　　tumnale* 1 50
　　　Standard: .5% colchicine, by weight.
　　　Antipodagric, cathartic, emetic.
　　　See note 49. Dose, 1 to 5 minims.

166. Colocynth, *Citrullus Colocynthis* . . . 1 90
　　　Drastic purgative. Dose, 5 to 10 minims.

167. Coltsfoot, *Tussilago Farfara* 1 00
　　　Demulcent. Dose, 30 to 60 minims.

168. Columbo, *Jateorhiza palmata* 1 20
　　　Tonic. Dose, 10 to 30 minims.

169. Comfrey, *Symphytum officinale* . . . 1 00
　　　Demulcent, vulnerary.
　　　Dose, 60 to 120 minims.

Per lb.

170. Condurango, *Gonolobus Condurango* . . . $3 00
   -   Alterative. Dose, 30 minims. . . .

171. Conium fruit, assayed, *Conium maculatum.* 1 50
   Standard: .5% coniine, by titration.
   Sedative, narcotic. See note 49.
   Dose, 2 to 5 minims.

172. Conium leaves, *Conium maculatum* . . . 1 20
   Sedative, narcotic. Dose, 5 to 10 minims.

   Convallaria (297). See Lily-of-the-Valley 1 25

173. Coriander, *Coriandrum sativum* 1 20
   Carminative, stimulant. ·   1
   Dose, 10 to 30 minims.

174. Corn Ergot, *Ustilago Maydis* . . . 1 75
   Parturient, emmenagogue. See note 34.
   Dose, 15 to 30 minims.

175. Corn-silk, *Zea Mays* 1 50
   Diuretic, lithontriptic, demulcent.
   See note 35. Dose, 20 to 30 minims.

176. Corydalis, *Dicentra Canadensis*. . . . 1 80
   Tonic, diuretic, alterative.
   Dose, 15 to 30 minims.

177. Corydalis Comp., for N. F. Elixir. . . . 1 80
   100 Cc. represent— 1 fl. oz. represents— [
   ...24. Gm. Corydalis . . . . . 109 grs.
   24. Gm. Stillingia . . . . . 109 grs.
   12. Gm. Xanthoxylum. . . . 55 grs.
   36. Gm. Iris. . . . . 164 grs.
   Four times as strong as the Elixir.
   See note 50.

178. Coto bark, true. 4 00
   Intestinal astringent and antiseptic.
   See note 36. Dose, 5 to 15 minims.

179. Cotton-root bark, *Gossypium herbaceum* . 1 25
   Emmenagogue, oxytocic.
   Dose, 15 to 60 minims.

180. Couch-grass, *Agropyrum repens* . . . 1 05
   Demulcent diuretic. See note 102.
   Dose, 30 to 120 minims. . . . .

   Cramp-bark (480).   See *Viburnum
   opulus* 1 00

   Cramp-bark Comp. (481). See Viburnum Comp. 1 20

181. Cranesbill, *Geranium maculatum* . . 1 05
   Tonic, astringent. Dose, 20 to 40 minims.

Per lb.

Crataegus (244). See Hawthorn berries $2 00

182. Crawley-root, *Corallorhiza odontorhiza* . . 4 50
     Sedative, diaphoretic.
     Dose, 15 to 30 minims.

183. Cubeb, *Piper Cubeba* 2 50
     Stimulant, diuretic, carminative.
     Dose, 10 to 30 minims.

184. Culver's-root, *Veronica Virginica* . . . 1 20
     Alterative, cholagogue cathartic.
     Dose, 20 to 60 minims.

Cusso (281). See Kousso 2 40

Cypripedium (284). See Ladies'-slipper. 2 00

185. Damiana, *Turnera diffusa* 2 40
     Bitter tonic, aphrodisiac. See note 37.
     Dose, 30 to 60 minims.

186. Damiana Comp. 2 50
     100 Cc. represent— 1 fl. oz. represents—
         48. Gm. Damiana . . . . . . 219 grs.
         48. Gm. Coca . . . . 219 grs.
         4. Gm. Nux Vomica . . . . 18 grs.
     Cerebro-spinal stimulant, tonic, aphrodisiac.
     See note 38. Dose, 60 minims.

187. Dandelion, *Taraxacum officinale.* . . . 1 25
     Deobstruent, tonic. Dose, 30 to 120 minims.

188. Dandelion Comp. 1 25
     100 Cc. represent— 1 fl. oz. represents—
         81.25 Gm. Dandelion . . . . 370 grs.
         6.25 Gm. Conium leaves . . . 29 grs.
         12.50 Gm. Podophyllum . . . 57 grs.
     Cholagogue, laxative.
     Dose, 30 to 60 minims.

189. Dandelion and Senna. 1 20
     100 Cc. represent— 1 fl oz. represents—
         50. Gm. Dandelion . . . . . 228 grs.
         50. Gm. Senna . . . . 228 grs.
     Cholagogue, laxative.
     Dose, 30 to 60 minims.

190. Deer-tongue, *Trilisa odoratissima* . . . 1 20
     Aromatic stimulant. Dose, 30 to 60 minims.

191. Digitalis, *Digitalis purpurea* 1 25
     Cardiac tonic, diuretic. See note 40.
     Dose, 1 to 3 minims.

192. Dill, *Anethum graveolens* 2 25
     Stimulant, carminative, stomachic.
     Dose, 10 to 30 minims.

Per lb.

193. Dogwood, *Cornus Florida* $1 00
    Tonic. Dose, 10 to 30 minims.
    Dover's Powder (264). See Ipecac and
      Opium 2 40

194. Duboisia leaves, *Duboisia myaporoides* . . 4 00
    Diuretic, sedative, narcotic.
    See note 42. Dose, 1 to 5 minims.
    Dulcamara (58): See Bittersweet 1 20

195. Dwarf Elder, *Aralia hispida*. 1 00
    Diuretic. Dose, 60 to 120 minims.

196. Echinacea, *Echinacea angustifolia* . . . 2 00
    Alterative. See note 43.
    Dose, 10 to 20 minims.

197. Elder flowers, *Sambucus Canadensis* . . 1 00
    Carminative, diaphoretic.
    Dose, 30 to 60 minims.

198. Elecampane, *Inula Helenium* 1 00
    Diaphoretic, diuretic, expectorant.
    Dose, 20 to 60 minims.

199. Embelia, *Embelia Ribes.* 2 50
    Taenicide. See note 44.
    Dose, 1 to 4 fluidrachms.

200. Ephedra, *Ephedra antisyphilitica* . . . 3 50
    Antisyphilitic. See note 45.
    Dose, 60 to 120 minims.

201. Ergot, *Claviceps purpurea.* 1 80
    Ecbolic, emmenagogue, hemostatic.
    See note 46. Dose, 30 to 60 minims.
    Eriodictyon (521). See Yerba Santa . . 2 50

202. *Eschscholtzia Californica,* California
      Poppy 3 00
    Sedative, soporific, analgesic. See note 47.
    Dose, 15 to 30 minims.

203. Eucalyptus, *Eucalyptus globulus* 1 15
    Febrifuge, astringent, antiseptic.
    Dose, 15 to 60 minims.
    Euonymus (484). See Wahoo 1 50
    Eupatorium (79). See Boneset 1 00

204. Euphorbia, *Euphorbia pilulifera.* . . . 4 00
    Antasthmatic. See note 48.
    Dose, 15 to 60 minims.

205. European Elder bark, *Sambucus nigra* . 1 80
    Diuretic, alterative.
    Dose, 15 to 60 minims.

	Per lb.
206. Evening Primrose, *Oenothera biennis*	$1 50
Astringent, sedative.	
Dose; 30 to 60 minims.	
207. False Bittersweet, *Celastrus scandens*.	1 05
Diaphoretic, diuretic, emetic.	
Dose; 60 to 120 minims.	
False Hellebore (6). See *Adonis vernalis*.	3 00
208. False Unicorn, *Chamaelirium luteum*	1 80
Tonic, diuretic, anthelmintic.	
Dose; 15 to 60 minims.	
209. Fennel, *Foeniculum capillaceum*	1 20
Carminative, stomachic.	
Dose, 15 to 30 minims.	
210. Feverfew, *Matricaria Parthenium*	1 00
Carminative, emmenagogue, tonic.	
Dose, 60 to 120 minims.	
Figwort (119). See Carpenter's-square.	1 80
211. Fireweed, *Erechthites hieracifolia*	1 00
Tonic, astringent. Dose, 30 to 60 minims.	
Fishberries (159). See Cocculus Indicus.	1 20
212. Five-flowered Gentian, *Gentiana quinque-*	
flora	2 50
Tonic, antiperiodic.	
Dose, 10 to 30 minims.	
213. Fleabane, *Erigeron Canadense.*	1 20
Diuretic, stomachic. Dose, 30 to 60 minims.	
214. Flowering Spurge, *Euphorbia corollata*	1 80
Expectorant, laxative, diaphoretic, emetic.	
Dose, 5 to 20 minims.	
Foeniculum (209). See Fennel	1 20
Foxglove (191). See Digitalis	1 25
Frangula (92). See Buckthorn bark	1 00
215. *nanthus Virginica*.	1 50
Diuretic, aperient. Dose, 5 to 20 minims.	
216. Frostwort, *Helianthemum Canadense* .	1 00
Alterative, tonic, astringent.	
Dose, 30 to 60 minims.	
217. Galanga, *Alpinia officinarum*	1 80
Carminative, stomachic.	
Dose, 15 to 30 minims.	
218. Galls, *Quercus lusitanica*	1 50
Astringent. Dose, 10 to 20 minims.	

Per lb.

219. Garden Celandine, *Chelidonium majus* . $1 00
    Drastic cathartic.   Dose, 15 to 60 minims.

220. Garlic, *Allium sativum* 1 20
    Diuretic, expectorant.
    Dose, 30 to 60 minims.

221. Gelsemium, assayed, *Gelsemium semper-*
    virens 1 25
    Standard : .5% total alkaloids, by titration.
    Sedative, diaphoretic. nervine.
    See notes 49, 52.   Dose, 3 to 10 minims.

222. Gentian, *Gentiana lutea* 1 00
    Tonic.   Dose, 10 to 30 minims.

223. Gentian Comp., for U. S. P. Tincture .   1 15
    100 Cc. represent— 1 fl. oz. represents—
    53.33 Gm. Gentian. . . . . . 243 grs.
    21.33 Gm. Bitter Orange.   97 grs.
    5.34 Gm. Cardamom . . . . 24 grs.
    Five and one-third times as strong as the
    Tincture.   See note 50.

224. Gentian Comp., for B. P. Infusion .   1 15
    100 Cc. represent— 1 imp. pint represents—
    20. Gm. Gentian . . . . . . 4 ozs. 10 grs.
    20. Gm. Bitter Orange peel . 4 ozs. 10 grs.
    40. Gm. Fresh Lemon peel . 8 ozs.
    Sixteen times as strong as the Infusion.
    See note 50.

225. Gentian Comp., for B. P. Tincture ,   1 15
    100 Cc. represent— 1 imp. pint represents—
    60. Gm. Gentian . . . . . . 12 ozs.
    30. Gm. Bitter Orange peel . 6 ozs.
    10. Gm. Cardamom seeds . . . . 2 ozs.
    Eight times as strong as the Tincture.
    See note 50.

    Geranium (181).   See Cranesbill . . .   1 05

226. Ginger, *Zingiber officinale* 1 50
    Carminative, stimulant, rubefacient.
    Dose, 10 to 20 minims.

227. Ginger, soluble, for Syrup 1 00
    See note 50.

    Glycyrrhiza (291).   See Licorice . .   1 00

228. Golden Rod, *Solidago odora* 1 00
    Stimulant, carminative.
    Dose, 30 to 60 minims.

Per lb.

229. - Golden Seal, assayed, *Hydrastis Cana-*
densis. $2 50
    1   Standard: 2.75% total alkaloids, by titration.
       Tonic, deobstruent, alterative. See note 49.
       Dose, 10 to 40 minims.

230. Golden Seal, colorless . . . ., 1 50
       For local use. See note 54.

231. Golden Seal, non-alcoholic 1 75
       For local use. See note 55.

232. Gold Thread, *Coptis trifolia* 2 25
       Tonic. Dose, 15 to 60 minims.

233. Gravel Plant, *Epigaea repens* 1 00
       Diuretic, nephritic. Dose, 15 to 60 minims.

234. Green Osier, *Cornus circinata* 1 05
       Tonic. Dose, 10 to 30 minims.

235. Grindelia, *Grindelia robusta* and *G. squar-*
rosa 1 90
       Antasthmatic, tonic. Dose, 15 to 60 minims.

236. Grindelia Aromatic, for Elixir 2 00
       100 Cc. represent— 1 fl. oz. represents— .
          52.66 Gm. Grindelia . . . . . . 240 grs.
          35.11 Gm. Licorice. . . . . . . 160 grs.
          8.77 Gm. Aromatics . . . . . 40 grs.
       Four times as strong as the Elixir.
       See note 50.

237. Grindelia Comp . . . ., . 2 00
       100 Cc. represent— 1 fl. oz. represents—
          66.67 Gm. Grindelia . . . . . . 304 grs.
          16.67 Gm. Senna. . . . . 76 grs.
          16.66 Gm. Rhubarb . . . . . . 76 grs.
       Antasthmatic, laxative.
       Dose, 30 to 60 minims.

238. Guaco root, *Agave planifolia* 2 40
       Febrifuge, anthelmintic, alterative.
       See note 56. Dose, 15 to 30 minims.

239. Guaiac resin, *Guaiacum officinale* . . . 1 50
       Stimulant, diaphoretic, alterative.
       See note 51. Dose, 10 to 30 minims.

240. Guaiac wood, *Guaiacum officinale* . . . 1 00
       Stimulant, diaphoretic, alterative.
       Dose, 15 to 60 minims.

241. Guarana, assayed, *Paullinia Cupana* . . 5 00
       Standard: 4.% caffeine, by weight.
       Tonic, stimulant, nervine.
       See notes 49, 57. Dose, 15 to 60 minims.

Per lb.

242. Haircap Moss, *Polytrichum juniperum* : $1 20
    Diuretic. Dose, 60 to 120 minims.

Hamamelis (511). See Witch Hazel 1. 1 00

243. Hardhack, *Spiraea tomentosa* 1 00
    Astringent, tonic. Dose, 30 to 60 minims.

244. Hawthorn berries, *Crataegus oxyacantha* 2 00
    Cardiac tonic. Dose, 5 to 15 minims.

Hedeoma (353). See Pennyroyal . . . 1 00

Helonias (208). See False Unicorn . . 1 80

245. Helonias Comp., for Elixir 1 50
    100 Cc. represent—. 1 fl. oz. represents—
      52.67 Gm. Helonias . . . . . 240 grs.
      13.17 Gm. Buchu . . . . . . 60 grs.
      13.17 Gm. Trillium . . . . . 60 grs.
      13.17 Gm. Gentian . . . . . 60 grs.
      13.17 Gm. Hydrastis . . . . 60 grs.
    Twice as strong as the Elixir. See note 58.

246. Hemlock Spruce bark, *Tsuga Canadensis* 1 00
    Tonic, astringent.
    Dose, 15 to 60 minims.

247. Henbane, assayed, *Hyoscyamus niger* . . 1 25
    Standard : .12% alkaloids, by weight.
    Anodyne, mydriatic, narcotic.
    See note 49. Dose, 2 to 10 minims.

248. Hops, *Humulus Lupulus* 1 70
    Tonic, hypnotic, diuretic.
    Dose, 15 to 60 minims.

249. Horehound, *Marrubium vulgare* . . . 1 00
    Bitter tonic, deobstruent.
    Dose, 30 to 60 minims.

250. Horehound Comp., for Eclec. Disp. Syrup 1 25
    100 Cc. represent— 1 fl. oz. represents—
      17.55 Gm. Horehound. . . . . 80 grs.
      17.55 Gm. Red root . . . . . . 80 grs.
      17.55 Gm. Elecampane . . . . . 80 grs.
      17.55 Gm. Spikenard . . . . . 80 grs.
      17.55 Gm. Comfrey . . . . . . 80 grs.
      17.55 Gm. Wild Cherry . . . . . 80 grs.
      8.78 Gm. Sanguinaria . . . . 40 grs.
    Four times as strong as the Syrup.
    See note 50.

251. Horse-chestnut bark, *Aesculus Hippocasta-*
    num 1 20
    Tonic, astringent, antiperiodic.
    Dose, 30 to 120 minims.

Per lb.

252. Horse-chestnuts, *Aesculus Hippocastanum.* $1 50
    Tonic, antiperiodic, antispasmodic.
    Dose, 5 to 15 minims.
253. Horse-nettle berries, *Solanum Carolinense* 2 00
    Antispasmodic, sedative. See note 59.
    Dose, 5 to 15 minims.
254. Horse-nettle root, *Solanum Carolinense* 2 00
    Antispasmodic, sedative. See note 59.
    Dose, 10 to 30 minims.
255. Horse-radish, *Cochlearia Armoracia* 1 20
    Stimulant, rubefacient.
    Dose, 60 to 120 minims.
    Humulus (248). See Hops 1 70
256. Hydrangea, *Hydrangea arborescens* . . 1 05
    Diuretic, antilithic. Dose, 30 to 60 minims.
    Hydrastis, assayed (229). See Golden
    Seal . . . . 1. . 2 50
    Standard: 2.75% total alkaloids, by titration.
    Hyoscyamus, assayed (247). See Hen-
    bane 1 25
    Standard: .12% alkaloids, by weight.
257. Hyssop, *Hyssopus officinalis* 1 00
    Carminative, sudorific.
    Dose, 15 to 60 minims.
258. Hysterionica, *Haplopappus Baylahuen* . 2 00
    Stomachic, intestinal antiseptic.
    See note 62. Dose, 5 to 15 minims.
259. Ignatia, assayed, *Strychnos Ignatia* . . 2 10
    Standard: 2.% alkaloids, by titration.
    Tonic, spinal nervine. See note 49.
    Dose, 1 to 4 minims.
260. Indian Black-root, *Pterocaulon pycnos-
    tachyum* 1 80
    Alterative. Dose, 15 to 30 minims.
    Indian Hemp (110). See Cannabis Indica 1 50
    See note 63.
261. Indian Physic, *Gillenia trifoliata* . . 1 00
    Mild emetic, cathartic.
    Dose, 15 to 45 minims.
262. Indian Turnip, *Arisaema triphyllum* . 1 20
    Expectorant, diaphoretic.
    Dose, 5 to 15 minims.
    Inula (198). See Elecampane 1 00

Per lb.

263. Ipecac, assayed, *Cephaëlis Ipecacuanha* . $5 00
  Standard : 2.% emetine, by titration.
  Expectorant, diaphoretic, emetic.
  See note 49.   Dose, 3 to 30 minims.

264. Ipecac and Opium, U. S. P. Tincture . . 2 40
  100 Cc. represent— 1 fl. oz. represents—
    10. Gm. Ipecac . . . . 45.8 grs.
    10. Gm. Opium . . . . 45.8 grs.
  Diaphoretic, sedative.   See note 41.
  Dose, 5 to 10 minims.

265. Ipecac and Senega 2 40
  100 Cc. represent— 1 fl. oz. represents—
    36.21 Gm. Ipecac . . . . 165 grs.
    69.12 Gm. Senega . . . . 315 grs.
  Expectorant, diaphoretic, emetic.
  See note 50.   Dose, 10 to 40 minims.

Iris (77).   See Blue Flag 1 05

266. Ironwood, *Ostrya Virginica* 1 50
  Antiperiodic, tonic.   Dose, 30 to 60 minims.

267. Jaborandi, assayed, *Pilocarpus Selloanus.* 2 50
  Standard : .6% pilocarpine, by weight.
  Sialagogue, diaphoretic.   See note 49.
  Dose, 15 to 60 minims.

268. Jalap, assayed, *Ipomoea Jalapa* . . . 2 00
  Standard : 12.% resin.
  Hydragogue cathartic, diuretic.
  See note 49.   Dose, 10 to 30 minims.

269. Jamaica Dogwood, *Piscidia Erythrina* . 3 00
  Mild soporific.   See note 64.
  Dose, 15 to 45 minims.

270. Jambul seed, *Eugenia Jambolana* . . . 5 00
  Astringent.   See note 65.
  Dose, 5 to 10 minims.

271. Jersey Tea, *Ceanothus Americanus* . . . 1 20
  Astringent, expectorant.
  Dose, 10 to 30 minims.

272. Johnswort, *Hypericum perforatum* . . . 1 00
  Stimulant, diuretic, astringent.
  Dose, 30 to 60 minims.

273. Judas Tree, *Cercis Canadensis* 3 00
  Astringent.   Dose, 30 to 60 minims.

Juglans (97).   See Butternut 1 00

274. Juniper berries, *Juniperus communis* . . 1 00
  Stimulant, diuretic, carminative.
  Dose, 15 to 60 minims.

Per lb.

275. Jurubeba, *Solanum paniculatum* $3 00
　　　Tonic, deobstruent.
　　　See note 66. Dose, 5 to 30 minims.

276. Kamala, *Mallotus Philippinensis* 3 00
　　　Taenifuge, purgative. See note 67.
　　　Dose, 60 to 120 minims.

277. Kavakava, *Piper methysticum* 2 00
　　　Tonic, nervine, diuretic. See note 68.
　　　Dose, 30 to 60 minims.

278. Kino, U. S. P. Tincture, *Pterocarpus
　　　Marsupium* 2 00
　　　Astringent. See note 69.
　　　Dose, 30 to 120 minims.

279. Kola, assayed, *Sterculia acuminata*. . . 2 00
　　　Standard: 1.% caffeine, by weight.
　　　Cerebro-spinal stimulant, diuretic, cardiac
　　　　　tonic, nervine and conservator of energy.
　　　See notes 49, 70. Dose, 10 to 60 minims.

280. Kola, fresh nut, *Sterculia acuminata* . . 2 00
　　　See note 70. Dose, 15 to 30 minims.

281. Kousso, *Hagenia Abyssinica*. 2.40
　　　Taenifuge. See note 71.
　　　Dose, 2 to 4 fluidrachms.

　　　Krameria (386). See Rhatany 1 10

282. Lactucarium, *Lactuca virosa* 5 00
　　　Anodyne, hypnotic, sedative.
　　　Dose, 5 to 60 minims.

283. Ladies' Bedstraw, Yellow, *Galium verum* 1 50
　　　Diuretic, alterative. Dose, 15 to 60 minims.

284. Ladies'-slipper, *Cypripedium pubescens* . 2 00
　　　Diaphoretic, antispasmodic.
　　　Dose, 8 to 15 minims.

　　　Lappa (95). See Burdock root 1 00

285. Larkspur seed, *Delphinium consolida* . . 3 50
　　　Emetic, cathartic, diuretic.
　　　Dose, ½ to 3 minims.

286. Lavender, garden, *Lavandula vera* . . 1 50
　　　Stimulant, carminative.
　　　Dose, 15 to 30 minims.

Per lb.

287. Lavender Comp., for U. S. P. 'Tincture . $1 60
100 Cc. represent— 1 fl. oz. represents—
    6.4 Cc., Oil Lavender flowers . 31 m.
    1.6 Cc. Oil Rosemary . . . . 8 m.
    16. Gm. Cassia Cinnamon . . 73 grs.
    4. Gm. Cloves . . . . 18 grs.
    8. Gm. Nutmeg . . . . . . 36 grs.
    8. Gm. Red Saunders . . . . 36 grs.
Eight times as strong as the Tincture.
See note 50.

288. Lavender Comp., for B. P. Tincture . . 1 60
100 Cc. represent— 1 imp. pint represents—
    1.87 Cc. Oil Lavender . . . 3 fl. drachms.
    21 Cc. Oil Rosemary . . 20 m.
    3.43 Gm. Cinnamon . . . 300 grs.
    3.43 Gm. Nutmeg . . . . 300 grs.
    6.85 Gm. Red Sandalwood 600 grs.
Four times as strong as the Tincture.
See note 50.

289. Lemon peel, *Citrus limonum* 1 20
For flavoring.

Leptandra (184). See Culver's-root . . 1 20

Lettuce, Wild (507). See Wild Lettuce 1 20

290. Levant Wormseed, *Artemisia pauciflora* . 1 50
Anthelmintic. Dose, 10 to 60 minims.

291. Licorice, *Glycyrrhiza glabra*. 1 00
Demulcent, expectorant.
Dose, 15 to 60 minims.

292. Licorice, Aromatic 1 20
For Quinine mixtures. See note 72.

293. Life-everlasting, *Gnaphalium polycepha-
lum* 1 00
Tonic, astringent, vulnerary.
Dose, 15 to 60 minims.

294. Life-root, *Senecio aureus* 1 00
Emmenagogue, vulnerary.
Dose, 30 to 60 minims.

295. Lily-of-the-Valley flowers, *Convallaria
majalis* 2 00
Cardiac tonic, diuretic. See note 73.
Dose, 10 to 30 minims.

296. Lily-of-the-Valley herb, *Convallaria ma-
jalis* 1 25
Cardiac tonic, diuretic. See note 73.
Dose, 10 to 30 minims.

30   FREDERICK STEARNS. & CO.'S

Per lb.

297. Lily-of-the-Valley root, *Convallaria ma-
jalis* $1 25
Cardiac tonic, diuretic. See note 73.
Dose, 10 to 30 minims.

298. Lippia Mexicana, Tincture, *Lippia dulcis* . 3 20
Demulcent, expectorant.
Dose, 30 to 60 minims.

Liquorice (291). See Licorice 1 00

299. Liverwort, *Anemone hepatica* 1 50
Demulcent, tonic. Dose, 30 to 120 minims.

300. Lobelia herb, *Lobelia inflata* 1 05
Expectorant, diaphoretic, antasthmatic.
Dose, 5 to 30 minims.

301. Lobelia seed, *Lobelia inflata*. 2 50
Expectorant, diaphoretic, antasthmatic.
Dose, 3 to 10 minims.

302. Lobelia Comp., for Eclec. Disp. Tincture . 1 05
100 Cc represent— 1 fl. oz. represents—
13.17 Gm. Lobelia herb . . . . 60 grs.
13.17 Gm. Blood-root . . . 1 . 60 grs.
13.17 Gm. Skunk Cabbage . . . 60 grs.
13.17 Gm. Wild Ginger . . . . 60 grs.
13.17 Gm. Pleurisy-root . . . . 60 grs.
Eight times as strong as the Tincture.
See note 50.

303. Logwood, *Haematoxylon Campechianum*. 1 00
Astringent. Dose, 60 to 120 minims.

304. Lovage, *Ligusticum levisticum* 1 50
Stimulant, carminative.
Dose, 10 to 30 minims.

305. Lungwort, *Pulmonaria officinalis* 1 00
Pectoral, demulcent.
Dose, 30 to 60 minims.

306. Lupulin 2 50
Tonic, hypnotic, diuretic.
Dose, 10 to 30 minims.

Lupulin Comp. (525) 1 80
100 Cc. represent— 1 fl. oz. represents—
33.33 Gm. Lupulin . . . . . 152 grs.
33.33 Gm. Wild Lettuce . . . 152 grs.
33.34 Gm. Scutellaria . . . . 152 grs.
Tonic, hypnotic, nervine. See note 50.
Dose, 20 to 60 minims.

307. Mace, *Myristica fragrans* 2 50
Carminative, stimulant.
Dose, 5 to 30 minims.

Per lb.

308. Male-fern, assayed, *Dryopteris Filix-mas.*$1 25
　　　Standard: '10.% oleoresin.'
　　　Taenifuge.　See notes 49, 75.
　　　. . Dose, 1/to/4 fluidrachms..

309. Manacá, *Brunfelsia Hopeana* ,　　　　　3 50
　　　Alterative, antirheumatic. See note 76.
　　　Dose, 15 to 60 minims.

310. Mandrake, assayed, *Podophyllum peltatum* 1 05
　　　Standard: . 4.% resin. . !
　　　Cathartic, cholagogue. .See note 49.
　　　Dose, 4 to 20 minims.

311. Mandrake Comp. 1 05
　　　100 Cc. represent—1 fl. oz. represents— 1
　　　37.50 Gm. Podophyllum, v1. 1171 grs.
　　　25　Gm. Leptandra . . . . . 114 grs.
　　　12.50 Gm. Jalap. . . . . . . 57 grs.
　　　25 '' Gm. Senna . . . . . 114 grs.
　　　Cholagogue, cathartic. 'See note 50.'
　　　Dose, 15 to 60 minims.

312. Mangosteen, *Garcinia Mangostana.* . . 3 00
　　　Astringent.　Dose, 15 to 60 minims.

313. Manzanito leaves, *Arctostaphylos glauca* . 2 40
　　　Astringent, tonic.　Dose, 20 to 60 minims.

　　　Marrubium (249). '' See Horehound' . . 1 00

314. Marshmallow, *Althaea officinalis* . . . 1 00
　　　Demulcent. Dose, 30 to 120 minims.

315. Marsh Rosemary, *Statice Limonium* var.
　　　' *Caroliniana* 1 00
　　　Astringent, tonic ; topical application.
　　　Dose, 15 to 60 minims.

316. Masterwort, *Heracleum lanatum* 1 20
　　　.- Stimulant, antispasmodic, carminative.
　　　. Dose, 60 to 120 minims.

317. Matico, *Piper angustifolium* 1 80
　　　. Stimulating diuretic, and expectorant.
　　　Dose, 15 to 60 minims. .

　　　Matricaria (138).　See Chamomile, Ger-
　　　man　. . . . . 1 20

　　　Melissa (37).　See Balm 1 20

　　　Menispermum (517).　See Yellow Parilla 1 20

318. Mezereum, *Daphne Mezereum* . 1 . . . . 1 50
　　　Alterative, stimulant, diuretic.
　　　Dose, 2 to 10 minims.

　　　Milkweed (425). . See Silkweed . 1. . 1 20

Per lb.

319. Mistletoe, *Phoradendron flavescens* $1 20
    Laxative, oxytocic, antispasmodic.
    Dose, 15 to 60 minims.

Mitchella (437). See Squaw-vine . . .   1 20

320. Mitchella Comp., for Eclec. Disp. Syrup .   1 50
    100 Cc. represent— 1 fl. oz. represents—
      52.67 Gm. Partridgeberry . . . 240 grs.
      13.17 Gm. Helonias root . . . . 60 grs.
      13.17 Gm. High Cranberry bark 60 grs.
      13.17 Gm. Blue Cohosh . . . . 60 grs.
    Four times as strong as the Syrup.
    See note 58.

321. Motherwort, *Leonurus cardiaca* 1 05
    Emmenagogue, antispasmodic.
    Dose, 30 to 60 minims.

322. Mountain Laurel, *Kalmia latifolia* . . .   1 20
    Cardiac sedative, astringent, alterative.
    Dose, 15 to 30 minims.

323. Mountain Mint, *Pycnanthemum montanum* 1 00
    Carminative, stimulant.
    Dose, 15 to 60 minims.

Mountain Rush (200). See *Ephedra antisyphilitica* 3 50

Mountain Sage (424). See Sierra Salvia. 2 50

324. Mugwort, *Artemisia vulgaris* 1 00
    Anthelmintic, vulnerary, emmenagogue.
    Dose, 20 to 60 minims.

325. Mugwort Comp 1 00
    100 Cc. represent— 1 fl. oz. represents—
      85.58 Gm. Mugwort . . . . . 390 grs.
      6.58 Gm. Bitter Orange peel . 30 grs.
      6.58 Gm. Licorice . . . . . 30 grs.
      3.29 Gm. Spicewood berries. . 15 grs.
    Alterative, tonic, emmenagogue.
    See note 50. Dose, 15 to 60 minims.

326. Muirapuama, *Liriosma ovata* 4 00
    Aphrodisiac. See note 77.
    Dose, 15 to 60 minims.

327. Mullein leaves, *Verbascum Thapsus* .   1 00
    Demulcent, anodyne, pectoral.
    Dose, 1 to 4 fluidrachms.

328. Mullein root, *Verbascum Thapsus* . . .   1 25
    Demulcent, anodyne, pectoral.
    Dose, 30 to 120 minims.

	Per lb.

329. Musk-root, *Ferula Sambul* . . . . . : $3 25
  Stimulant, tonic, nervine.
  Dose, 10 to 20 minims.

330. *Mutisia viciaefolia*, Chinchirocoma . . .   3 00
  Antispasmodic, expectorant, cardiac tonic.
  See note 78.   Dose, 15 to 60 minims.

331. Myrrh, *Commiphora Myrrha* 2 10
  Stimulant, expectorant, emmenagogue, vul-
    nerary. See note 51.
    Dose, 5 to 30 minims.
  Myrrh and Capsicum (113). See Capsi-
    cum and Myrrh 2 00

332. Nettle root, *Urtica dioica* . . . . . . :   1 20
  Diuretic, astringent, tonic.
  Dose, 20 to 40 minims.

333. Newbouldia, *Newbouldia laevis* 3 00
  Astringent.   See note 79.
  Dose, 15 to 60 minims.
  New Jersey Tea (271). See Jersey Tea .   1 20
  Night-blooming Cereus (135).   See *Cer-
    eus grandiflorus* 4 00
  Nutgalls (218).   See Galls 1 50

334. Nutmeg, *Myristica fragrans* 3 00
  Aromatic, stomachic.   Dose, 10 to 30 minims.

335. Nux Vomica, assayed, *Strychnos Nux
    Vomica* 1 20
  Standard: 1.5% total alkaloids, by titration.
  Tonic, spinant.   See note 49.
  Dose, 1 to 4 minims.

336. Opium, concentrated, assayed, for U. S. P.
    and B. P. Tinctures 5 50
  Standard: 5.6% morphine.
  Four times as strong as the U. S. P.
    Tincture.   See notes 49, 82.

337. Opium, aqueous and deodorized, assayed,
    for U. S. P. Tincture 5 50
  Standard: 5.6% morphine.
  Four times as strong as Tincture Deod.
    Opium, U. S. P.   See notes 49, 82.
  Opium, camphorated, for B. P. Tincture
    (106).   See Camphor Comp. 1 50

Per lb.

338. Opium, camphorated, for U. S. P. Tincture.$1 75
100 Cc. represent— 1 fl. oz. represents—
    3.2 Gm. Opium . . . . 14.6 grs.
    3.2 Gm. Camphor . . . . . 14.6 grs.
    3.2 Gm. Benzoic acid . . . . 14.6 grs.
    3.2 Cc. Oil Anise . . . . . 15.3 m.
    32. Cc. Glycerin . . . . . 153.6 m.
Eight times as strong as the Tincture.
See notes 50, 82.
Opium, Tinctures. See Tincture list.

339. Orange peel, bitter, *Citrus vulgaris* ... 1 40
Tonic, carminative, stomachic.
Dose, 15 to 30 minims.

340. Orange peel, sweet, *Citrus aurantium* . 1 50
Tonic, carminative, stomachic.
Dose, 15 to 30 minims.

341. Orange Comp., for flavoring 1 80
100 Cc. represent— 1 fl. oz. represents—
    52.67 Gm. Sweet Orange peel . 240 grs.
    13.17 Gm. Licorice root . . . . 60 grs.
    13.17 Gm. Orris root . . . . . 60 grs.
    6.58 Gm. Cassia . . . . 30 grs.
    6.58 Gm. Mace . . . . 30 grs.
    3.29 Gm. Cardamom . . . . . 15 grs.
    3.29 Gm. Anise . . . . 15 grs.
    3.29 Gm. Coriander . . . . . 15 grs.
    3.29 Gm. Caraway . . . . . . 15 grs.
See note 50.

342. Orris root, *Iris florentina* 1 50
Cathartic, diuretic; also for flavoring.
Dose, 5 to 15 minims.

343. Ox-eye Daisy, *Leucanthemum vulgare* . 2 50
Anhydrotic in phthisis.

344. Pansy, *Viola tricolor* 1 20
Alterative, emollient.
Dose, 30 to 120 minims.

345. Papaw seed, *Asimina triloba* 1 50
Emetic, antasthmatic.
Dose, 10 to 30 minims.

346. Paraguay Tea, *Ilex Paraguayensis* . . . 1 80
Diuretic, stimulant. See note 84.
Dose, 30 to 60 minims.

347. Pareira Brava, *Chondodendron tomentosum* 1 50
Diuretic, tonic, aperient.
Dose, 15 to 60 minims.

Per lb.

348. Parsley fruit, *Petroselinum sativum*. . . $1 35
     Emmenagogue, laxative, diuretic:
     Dose, 30 to 60 minims,

349. Parsley root, *Petroselinum sativum* . . .   1 20
     Emmenagogue, laxative, diuretic.
     Dose, 30 to 60 minims.

     Partridgeberry Comp. (320). See Mitch-
       ella Comp. 1 50

     Pasque flower (376). See Pulsatilla . .   1 50

350. *Passiflora incarnata*, Passion flower .   2 00
     Nerve sedative. See note 85.
     Dose, 10 to 20 minims.

351. Peach-tree leaves, *Prunus Persica* . . .   1 25
     Mild sedative. Dose, 15 to 30 minims.

352. Pellitory, *Anacyclus Pyrethrum* 1 20
     Irritant; used externally.

     Pencil-flower (448). See *Stylosanthes
       elatior* 4 00

353. Pennyroyal, *Hedeoma pulegioides*. . . .   1 00
     Carminative, diaphoretic, emmenagogue.
     Dose, 15 to 60 minims.

354. Peony, *Paeonia officinalis*. 2 40
     Nervine, emmenagogue.
     Dose, 15 to 60 minims:

     Pepo (377). See Pumpkin seed. 1 80
     Pepper, Black (67). See Black Pepper .   2 00

355. Peppermint, *Mentha piperita* 1 50
     Carminative. Dose, 15 to 60 minims.

356. Persimmon bark, *Diospyros Virginiana* .   1 20
     Astringent. Dose, 15 to 60 minims.

357. Persimmon, green fruit 2 00
     Astringent. Dose, 30 to 120 minims.

     Physostigma, assayed (101). See Calabar
       bean 3 60
     Standard: .25% physostigmine, by weight.

     Phytolacca fruit (369). See Poke berries.   1 50
     Phytolacca root (370). See Poke root .   1 00

358. Pichi, *Fabiana imbricata* 4 00
     Terebinthinate diuretic. See note 87.
     Dose, 10 to 40 minims.

359. Pimenta, *Pimenta officinalis* 1 25
     Stomachic, carminative.
     Dose, 10 to 40 minims.

Per lb.

360. Pink-root, *Spigelia Marilandica* $1 50
Anthelmintic. b Dose, 15 to 60 minims. .

361. Pink-root Comp. 1 50
100 Cc. represent— 1 fl. oz. represents—
18.09 Gm. Spigelia . . . . . . 82.5 grs.
18.09 Gm. Mandrake . . . . . 82.5 grs.
18.09 Gm. Canadian Hemp . . 82.5 grs.
18.09 Gm. Bitter-root . . . . 82.5 grs.
32.90 Gm. Balmony . . . . . 150. grs.
Anthelmintic, cathartic, diaphoretic.
See note 50. Dose, 15 to 60 minims.

362. Pink-root and Senna 1 50
100 Cc. represent— 1 fl. oz. represents—
65.83 Gm. Spigelia . . . . . 300 grs.
39.50 Gm. Senna . . . . . . 180 grs.
.31 Cc. Oil Anise . . . . 1 1-2 m.
.31 Cc. Oil Caraway . . . . 1 1-2 m.
Anthelmintic, cathartic. See note 50.
Dose, 1 to 3 fluidrachms.

Pinus Canadensis (246). See Hemlock
Spruce 1 00

363. Pinus Canadensis Comp., colorless . . . 1 00
For external use. See note 50.

Pinus Comp., for Syrup (496). See
White Pine Comp. 2 00

364. Pipsissewa, *Chimaphila umbellata* . . . 1 00
Astringent, diuretic, nephritic.
Dose, 15 to 60 minims.

365. Pitcher-plant root, *Sarracenia purpurea* . 1 50
Tonic, diuretic. Dose, 15 to 30 minims.

366. Plantain leaves, *Plantago major* . . . 1 00
Alterative, diuretic, refrigerant.
Dose, 30 to 60 minims.

367. Pleurisy-root, *Asclepias tuberosa* . . . 1 20
Diaphoretic, expectorant.
Dose, 30 to 60 minims.

Podophyllum, assayed (310). See Mandrake. 1 05
Standard: 4.% resin.

368. Poison oak, *Rhus radicans* 1 80
Irritant, narcotic. Dose, 5 to 15 minims.

369. Poke berries, *Phytolacca decandra* . . 1 50
Alterative, laxative, emetic.
Dose, 10 to 60 minims.

Per lb.

370. Poke root, *Phytolacca decandra* . . . $1 00
  Alterative, cathartic, emetic..
  Dose, 5 to 30 minims.

371. Pomegranate-root bark, *Punica granatum* 1 25
  Taenifuge. · Dose, 30 to 120 minims.

372. Poppy heads, *Papaver somniferum* . . . 1 05
  Mild narcotic. Dose, 15 to 45 minims.

373. Prickly-ash bark, *Xanthoxylum Ameri-
  canum* 1 10
  Alterative, stimulant, sialagogue.
  Dose, 15 to 30 minims.

374. Prickly-ash berries, *Xanthoxylum Ameri-
  canum* 2 50
  Alterative, tonic, stimulant.
  Dose, 5 to 10 minims.

375. Pride of China, *Melia Azedarach* . . . 2·00
  Anthelmintic, cathartic.
  Dose, 15 to 60 minims.

  Prunus Virginiana, (502). See Wild
  Cherry 1 05

376. Pulsatilla, *Anemone Pulsatilla* 1 50
  Cardiac sedative, emmenagogue.
  Dose, 1 to 5 minims.

377. Pumpkin seed, *Cucurbita Pepo* 1 80
  Taenifuge. Dose, 1 to 2 fl. ozs.

  Pyrethrum (352). See Pellitory 1 20

378. Quassia, *Picraena excelsa* 1 00
  Tonic. Dose, 15 to 30 minims.

379. Quebracho, *Aspidosperma Quebracho-
  blanco* 2 50
  Standard: 1.% alkaloids, by weight.
  Tonic, antasthmatic. · See notes 49, 93.
  Dose, 15 to 40 minims.

380. Queen-of-the-Meadow, *Eupatorium pur-
  pureum* 1 00
  Diuretic, tonic, astringent.
  Dose, 30 to 60 minims.

  Quillaja (428). See Soap-tree bark . . 1 20

381. Quinine-flower, *Sabbatia Elliottii* . . . 2 25
  Tonic, antiperiodic. Dose, 60 minims.

382. Raspberry leaves, *Rubus idaeus* 1 00
  Astringent. Dose, 20 to 60 minims.

  Red Clover (155). See Clover, blossoms, Red 1 20

Per lb.

Red Clover ·Comp.· (469). See Trifolium Comp. $2 00

383. Red Gum, *Eucalyptus rostrata* 2 50
Astringent, antiseptic, styptic.
Dose, 30 to 60 minims.

384. Red Osier bark, *Cornus sericea* 1 05
Tonic. Dose, 15 to 60 minims.
Red Rose (395). See Rose, Red . . . 2 50

385. Red Saunders, *Pterocarpus santalinus* . 1 65
Coloring agent.
Rhamnus Purshiana (121). See Cascara
Sagrada 1 50

386. Rhatany, *Krameria triandra* 1 10
Astringent. Dose, 15 to 30 minims.

387. Rhubarb, *Rheum officinale* 2 25
Tonic, purgative, astringent.
Dose, 10 to 30 minims.

388. Rhubarb, Aromatic, for U. S. P. Tincture
and Syrup 2 00
100 Cc. represent— 1 fl. oz. represents—
60. Gm. Rhubarb . . . . 273 grs.
12. Gm. Cassia Cinnamon . . . 55 grs.
12. Gm. Cloves . . . . 55 grs.
6. Gm. Nutmeg . . . . 27 grs
Three times as strong as the Tincture.
Twenty times as strong as the Syrup.
See note 50.

389. Rhubarb Comp., for Syrup of Rhubarb
and Potash Comp. 1 80
100 Cc. represent— 1 fl. oz. represents—
26.33 Gm. Rhubarb . . . . . 120 grs.
13.17 Gm. Cinnamon . . . . . 60 grs.
13.17 Gm. Golden Seal . . . 60 grs.
.52 Cc. Oil Peppermint . . 2½ m.
Eight times as strong as the Syrup.
See note 50.

390. Rhubarb and Senna, for Tincture . . . 1 80
100 Cc. represent— 1 fl. oz. represents—
13.17 Gm. Rhubarb . . . . . 60 grs.
3.29 Gm. Senna . . . . . . 15 grs.
1.65 Gm. Coriander . . . . 7½ grs.
1.65 Gm. Fennel . . . . . 7½ grs.
.82 Gm. Licorice . . . . . 3¾ grs.
65.83 Gm. Raisins . . . . 300 grs.
Five and one-third times as strong as the
Tincture. See note 50.

Per lb.

391. Rhubarb, Sweet, for Sweet Tincture of
Rhubarb, U. S. P. $1 80
100 Cc. represent— 1 fl. oz. represents—
53.33 Gm. Rhubarb . . . . . . 243 grs.
21.33 Gm. Glycyrrhiza . . . . . 97 grs.
21.33 Gm. Anise. . . . . . . 97 grs.
5.33 Gm. Cardamom . . . . . 24 grs.
Five and one-third times, as strong as the
Tincture. See note 50.

392. Rhubarb, for B. P. Syrup 2 00
See note 50.

393. Rhubarb, for B. P. Tincture. 3 00
100 Cc. represent— 1 imp.pint represents—
80. Gm. Rhubarb root, ... 16 ozs.
10. Gm. Cardamom seeds . . . 2 ozs.
10. Gm. Coriander fruit . . . . 2 ozs.
10. Gm. Saffron . . . . 2 ozs.
Eight times as strong as the Tincture.
See note 50.

394. *Rhus aromatica*, Sweet Sumach 3 00
Diuretic. See note 94.
Dose, 5 to 30 minims.
Rhus glabra (450). ·See Sumach berries. 1 00
Rhus Toxicodendron (368). See Poison
Oak 1 80

395. Rose, red, *Rosa gallica* 2 50
Mild astringent. Dose, 15 to 60 minims.

396. Rosinweed, *Silphium laciniatum* 1 80
Expectorant, diuretic, febrifuge. .
Dose, 10 to 40 minims.
Rottlera (276). See Kamala 3 00
Rubus (61). See Blackberry-root bark . 1 00

397. Rue, *Ruta graveolens* 1 00
Emmenagogue, diaphoretic. .
Dose, 5 to 20 minims.
Rumex (515). See Yellow Dock . . . 1 10
Rumex Comp. (516). See Yellow Dock
Comp. 1 50
Sabadilla (136). See Cevadilla seed . . 1 80

398. *Sabbatia campestris* 2 40
Tonic, antiperiodic. Dose, 15 to 60 minims.
Safflower (21). See American Saffron . 2 40

399. Sage, *Salvia officinalis* 1 00
Tonic, astringent, vulnerary.
Dose, 15 to 60 minims.

| | Per lb. |

Sambucus (197). See Elder flowers . . $1 00

400. Sandalwood, *Santalum album* 3 00
　　Stimulating diuretic.
　　Dose, 60 to 120 minims,
Sanguinaria, assayed, (.74). See Blood-
root 1 05
　　Standard : 3.5% total alkaloids, by titration.

401. Sarsaparilla, *Smilax officinalis* 1 50
　　Alterative. Dose, 60 to 120 minims.

402. Sarsaparilla Comp. 1 50
　　100 Cc. represent— 1 ff. oz. represents—
　　　75. Gm. Sarsaparilla . . . . . 342 grs.
　　　12. Gm. Glycyrrhiza . . . . . 55 grs.
　　　10. Gm. Sassafras . . . . . . 45 grs.
　　　3. Gm. Mezereum . . . . . . 14 grs.
　　Alterative. See note 50.
　　Dose, 30 to 120 minims.

403. Sarsaparilla Comp., for U. S. P. Syrup . 1 50
　　100 Cc. represent— 1 fl. oz. represents—
　　　80. Gm. Sarsaparilla. . . . 365 grs.
　　　6. Gm. Glycyrrhiza. . . . 27 grs.
　　　6. Gm. Senna . . . . . 27 grs.
　　　.04 Cc. Oil Sassafras. . . . 0.2 m.
　　　.04 Cc. Oil Anise . . . . . 0.2 m.
　　　.04 Cc. Oil Gaultheria . . . 0.2 m.
　　Four times as strong as the Syrup.
　　See note 50.

404. Sarsaparilla Comp., for B. P. Decoction . 1 50
　　100 Cc. represent— 1 imp. pint represents—
　　　100. Gm. Jamaica Sarsaparilla . 20 ozs.
　　　10. Gm. Sassafras root . . . . 2 ozs.
　　　10. Gm. Guaiac wood . . . . . 2 ozs.
　　　10. Gm. Licorice root . . . . . 2 ozs.
　　　5. Gm. Mezereon bark . . . . 1 oz.
　　Eight times as strong as the Decoction.
　　See note 50.

405. Sarsaparilla and Dandelion 1 50
　　100 Cc. represent— 1 fl. oz. represents—
　　　50 Gm. Sarsaparilla . . . . . . 228 grs.
　　　50 Gm. Dandelion. . . . . . 228 grs.
　　Alterative, tonic, laxative.
　　See note 50. Dose, 30 to 60 minims.

406. Sassafras, *Sassafras variifolium* 1 00
　　Diuretic, astringent. Dose, 15 to 60 minims.

407. Savine, *Juniperus Sabina* 1 25
　　Emmenagogue, diuretic, vermifuge.
　　Dose, 5 to 15 minims.

Per lb.

408. Saw Palmetto, *Serenoa serrulata* $2 00
 Nutritive tonic, diuretic, genito-urinary vital-
 izer. See note 95.
 Dose, 30 to 60 minims. . . . . '

409. Saxifrage, *Pimpinella saxifraga* ·ı ı. } . 2 00
 Diaphoretic, diuretic. stomachic.
 : ¯ ı Dose, 15 to 30 minims. · . : . . . 1 . · ·

410. Scarlet Pimpernel, *Anagallis arvensis* . . 2 50
 Deobstruent, antirheumatic.
 Dose, 60 minims.

 Scilla (438). See Squill 1 00

 Scoparius (81). See Broom 1 05

411. Scopolia, *Scopolia Carniolica* 1 50
 Mydriatic, narcotic, anhydrotic.
 Dose, 1 to 3 minims.

412. Scouring Rush, *Equisetum hyemale* . . . 1 80
 Diuretic, astringent. Dose, 15 to 60 minims.

413. Scullcap, *Scutellaria lateriflora* 1 50
 Tonic, nervine, antispasmodic.
 Dose, 30 to 60 minims.

414. Scullcap Comp. 1 25
 100 Cc. represent— 1 fl. oz. represents—
 37.30 Gm. Scullcap . . . . . . 170 grs.
 30.72 Gm. Cypripedium . . . . 140 grs.
 18.65 Gm. Lettuce. . . . 85 grs.
 18.65 Gm. Hops. . . . . . . 85 grs.
 Tonic, anodyne, hypnotic.
 See note 50. Dose, 30 to 60 minims.

415. Senega, *Polygala Senega* 2 40
 Expectorant, emetic.
 Dose, 10 to 30 minims.

416. Senna, *Cassia angustifolia* and *Cassia
  acutifolia* 1 50
 Cathartic. See note 96.
 Dose, 30 to 120 minims.

417. Senna, aqueous 1 75
 See note 96.

418. Senna Comp., for N. F. Syrup 1 50
 100 Cc. represent— 1 fl. oz. represents—
 54. Gm. Senna . . . . 246 grs.
 14. Gm. Rhubarb . . . . . . 64 grs.
 14. Gm. Frangula . . . . . . 64 grs.
 1.6 Gm. Oil Gaultheria . . . . 8 m.
 Four times as strong as the Syrup.
 See note 50.

Per lb.

419. Senna and Jalap $1 50
  100 Cc. represent— 1 fl. oz. represents—
   50. Gm. Senna . . . . 228 grs.
   50. Gm. Jalap . . . . 228 grs.
  Cathartic. See note 50.
  Dose, 30 to 60 minims.

420. Senna pods, *Cassia angustifolia* 1 75
  Cathartic. See note 97.
  Dose, 30 to 60 minims.

421. Serpentaria, *Aristolochia Serpentaria* . . 1 75
  Tonic, aromatic, stimulant.
  Dose, 30 to 60 minims.

422. Sheep Sorrel, *Rumex Acetosella* 1 20
  Refrigerant, diuretic.
  Dose, 60 to 120 minims.

423. Shepherd's-purse, *Capsella Bursa-pastoris* 1 20
  Diuretic, tonic, stimulant.
  Dose, 15 to 60 minims.

424. Sierra Salvia, *Artemisia frigida* 2 50
  Anthelmintic, antiperiodic, diuretic.
  See note 98. Dose, 60 to 120 minims.

425. Silkweed, *Asclepias syriaca* 1 20
  Pectoral, diuretic, alterative.
  Dose, 15 to 60 minims.

426. Simaruba, *Simaruba officinalis* 1 80
  Tonic. Dose, 8 to 30 minims.

427. Skunk Cabbage, *Symplocarpus foetidus* . 1 00
  Antispasmodic, narcotic, nauseant..
  Dose, 10 to 20 minims.

  Smartweed (490). See Water Pepper . 1 00

428. Soap-tree bark, *Quillaja saponaria* . . . 1 20
  Diuretic, irritant, detergent.
  Dose, 15 to 30 minims.

429. Soapwort, *Saponaria officinalis.* 1 00
  Alterative, detergent. Dose, 15 to 60 minims.

430. Solidago, *Solidago virgaurea,* 1 00
  Diuretic, astringent, lithontriptic.
  Dose, 30 to 60 minims.

431. Solomon's Seal, *Polygonatum biflorum* . 1 00
  Mild astringent, vulnerary.
  Dose, 60 to 120 minims.

432. Sourwood leaves, *Oxydendron arboreum.* 2 50
  Diuretic, refrigerant.
  Dose, 30 to 120 minims.

433. Spanish Needles, *Bidens bipinnata* . . . 1 20
  Emmenagogue. Dose, 15 to 60 minims.

	Per lb.
434. Spearmint, *Mentha viridis* . . . . · · · $1 00	
Carminative. Dose, 15 to 60 minims.	
435. Spicebush bark, *Lindera Benzoin* . . . 1 20	
'Aromatic stimulant, diaphoretic.'	
Dose, 60 to 120 minims.	
Spicebush Comp. (325). See Mugwort	
Comp. 1 00	
Spigelia (360). See Pink-root. 1 50	
Spigelia and Senna (362). See Pink-root	
and Senna. 1 50	
436. Spikenard, *Aralia racemosa* 1 00	
Alterative. Dose, 30 to 60 minims.	
437. Squaw-vine, *Mitchella repens* 1 20	
Tonic, astringent, diuretic.	
Dose, 30 to 60 minims.	
Squaw-vine Comp. (320). See Mitchella	
Comp. 1 50	
438. Squill, *Urginea maritima* 1 00	
Expectorant, diuretic, emetic, irritant.	
Dose, 1 to 10 minims.	
439. Squill Comp., for U. S. P. Syrup . . . 1 75	
100 Cc. represent— 1 fl. oz. represents—	
42.67 Gm. Squill . . . . , 195.1 grs.	
42.67 Gm. Senega . . . 195. grs.	
1.066 Gm. Tartar Emetic . 4.86 grs.	
Five and one-third times as strong as the	
Syrup. See note 50.	
440. Squill, Acetic, for B. P. Syrup 1 00	
Eight times as strong as the Syrup,	
See note 50	
Star-grass (474). See Unicorn 1 80	
441. Stavesacre seed, *Delphinium Staphisagria*. 2 50	
External; parasiticide.	
442. *Stillingia sylvatica*, Queen's-root. . . . 1 50	
Alterative. Dose, 15 to 30 minims.	
443. Stillingia Comp., for N. F. Syrup . . 1 50	
100 Cc. represent— 1 fl. oz. represents—	
25. Gm. Stillingia . . . . . . 114 grs.	
25. Gm. Corydalis . . . . . 114 grs.	
12.5 Gm. Iris . . . . 57 grs.	
12.5 Gm. Sambucus . . . . . 57 grs.	
12.5 Gm. Chimaphila . . . . 57 grs.	
6.5 Gm. Coriander . . . . . . 30 grs.	
6. Gm. Xanthoxylum berries. 27 grs.	
Four times as strong as the Syrup.	
See note 50.	

Per lb.

444. Stone-root, *Collinsonia Canadensis* $1 05
Diuretic, irritant. Dose, 30 to 60 minims.

445. Stramonium leaves, assayed, *Datura Stra-*
monium 1 05
Standard, .35% total alkaloids, by weight.
Narcotic, mydriatic. See note 49.
Dose, 1 to 3 minims.

446. Stramonium seed, assayed, *Datura Stra-*
monium 1 05
Standard: .35% total alkaloids, by weight.
Narcotic, mydriatic. See note 49.
Dose, 1 to 2 minims.

447. Strophanthus, Tincture, *Strophanthus*
hispidus 1 50
Cardiac tonic. See note 99.
Dose, 2 to 5 minims.

448. *Stylosanthes elatior*, Pencil-flower 4 00
Uterine sedative. Dose, 10 to 20 minims.

449. Sumach bark, *Rhus glabra* 1 00
Tonic, astringent, antiseptic.
Dose, 30 to 120 minims.

450. Sumach berries, *Rhus glabra* 1 00
Tonic, astringent, antiseptic.
Dose, 60 to 120 minims.

Sumach, sweet (394). See *Rhus aro-*
matica 3 00

Sumbul (329). See Musk-root 3 25

451. Summer Savory, *Satureia hortensis* . . . 1 00
Carminative, emmenagogue.
Dose, 15 to 60 minims.

452. Sundew, *Drosera rotundifolia* 2 50
Pectoral. Dose, 5 to 15 minims.

453. Sunflower seed, *Helianthus annuus* . . 1 20
Diuretic, expectorant.
Dose, 60 to 120 minims.

454. Swamp Milkweed, *Asclepias incarnata* . 1 20
Emetic, cathartic. See note 63.
Dose, 15 to 45 minims.

455. Sweet Fern, *Comptonia asplenifolia* . . . 1 00
Stimulant, astringent.
Dose, 15 to 30 minims.

Sweet Flag (102). See Calamus . . . 1 25

Per lb.

456. Sweet-gum bark, *Liquidambar styraciflua* $2 50
    Mucilaginous astringent.
    Dose, 30 to 60 minims.
  Sweet Orange peel (340). See Orange
    peel, sweet 1 50

457. Tag Alder, *Alnus serrulata* 1 00
    Tonic, astringent, alterative.
    Dose, 30 to 60 minims.

458. Tamarac, *Larix Americana* 1 20
    Tonic, mild astringent.
    Dose, 30 to 120 minims.

459. Tansy, *Tanacetum vulgare* 1 00
    Emmenagogue, stimulant, anthelmintic.
    Dose, 15 to 60 minims.
  Taraxacum (187). See Dandelion . . . 1 25

460. Tea, *Camellia Thea* 1 80
    Stimulant, astringent.
    Dose, 30 to 60 minims.

461. Thimble-weed, *Rudbeckia laciniata*. . . 1 20
    Diuretic, tonic. Dose, 15 to 60 minims.

462. Thyme, *Thymus vulgaris* 1 00
    Carminative, antispasmodic.
    Dose, 30 to 60 minims.

463. Tobacco, *Nicotiana Tabacum* 1 20
    Narcotic, sedative, emetic.
    Dose, 1 to 5 minims.

464. Tolu, soluble, for Syrup. 1 00
    See note 50.

465. Tolu, for Tincture, *Toluifera Balsamum*. 2 00
    Stimulant, expectorant, vulnerary.
    See note 51.

466. Tonga, *Raphidophora vitiensis* 5 50
    Anti-neuralgic. Dose, 30 to 60 minims.

467. Tonka Bean, *Dipteryx odorata* 5 00
    For flavoring.

468. Tormentilla, *Potentilla Tormentilla* . . 1 20
    Tonic, astringent. Dose, 10 to 30 minims.
  Trailing Arbutus (233). See Gravel
    Plant 1 00

Per lb.

469. Trifolium Comp., for Syrup $2 00
    100 Cc. represent— 1 fl. oz. represents—
        28.07 Gm. Red Clover . . . . . 128 grs.
        14.03 Gm. Stillingia . . . . . . 64 grs.
        14.03 Gm. Burdock . . . . . . 64 grs.
        14.03 Gm. Poke-root. . . . . . 64 grs.
        14.03 Gm. Berberis Aquifolium. 64 grs.
        14.03 Gm. Cascara Amarga . . 64 grs.
        3.50 Gm. Prickly Ash. . . . . 16 grs.
        7.01 Gm. Potassium iodide . . 32 grs.
    Four times as strong as the Syrup. . . .
    See notes 50, 101.

    Triticum (180). See Couch-grass 1 05

470. Trumpet plant, *Sarracenia flava* 2 50
    Tonic, diuretic. Dose, 15 to 30 minims.

471. Tulip-tree bark, *Liriodendron tulipifera.* 1 00
    Tonic, febrifuge. Dose, 60 to 120 minims.

    Turkey-corn (176). See Corydalis . . 1 80

472. Turmeric, *Curcuma longa* 1 20
    Coloring agent.

473. Twin-leaf, *Jeffersonia diphylla* 1 35
    Diuretic, expectorant, antispasmodic.
    Dose, 15 to 60 minims.

474. Unicorn-root, *Aletris farinosa* 1 80
    Tonic, diuretic. Dose, 10 to 30 minims.

    Ustilago Maydis (174). See Corn Ergot. 1 75

475. Uva Ursi, *Arctostaphylos Uva Ursi* . . . 1 00
    Diuretic, tonic, astringent.
    Dose, 20 to 60 minims.

476. *Vaccinium crassifolium* 1 25
    Astringent, diuretic, nephritic.
    Dose, 30 to 60 minims.

477. Valerian, *Valeriana officinalis* 1 10
    Anodyne, antispasmodic, nervine.
    Dose, 30 to 60 minims.

478. *Veratrum viride*, assayed, American Hel-
    lebore 1 50
    Standard: 1.% total alkaloids, by weight.
    Cardiac and spinal depressant, emeto-cathar-
    tic. See note 49. Dose; 1 to 2 minims.

479. Vervain, Blue, *Verbena hastata* 1 00
    Tonic, sudorific, expectorant.
    Dose, 30 to 60 minims.

    Vervain, White (499). See White Ver-
    vain 2 50

	Per lb.

480. *Viburnum opulus*, Cramp bark $1 00
    Antispasmodic. Dose, 60 to 120 minims.

481. Viburnum Comp., for N. F. Elixir . . . 1 20
    100 Cc. represent— 1 fl. oz. represents—
    24 Gm. Viburnum Opulus . . . 109 grs.
    48 Gm. Trillium . . . . . . 219 grs.
    24 Gm. Aletris . . . . 109 grs.
    Three and one-fifth times as strong as the
    Elixir. See notes 50, 103.
    Viburnum prunifolium (65). See Black
        Haw 1 20

482. Virginia Stonecrop, *Penthorum sedoides* . 2 40
    Astringent, demulcent.
    Dose, 15 to 30 minims.

483. Wafer Ash, *Ptelea trifoliata* 1 10
    Tonic, antiperiodic. Dose, 15 to 30 minims.

484. Wahoo, *Euonymus atropurpureus* . . . 1 50
    Tonic, laxative, antiperiodic.
    Dose, 30 to 60 minims.
    Walnut leaves, Black (68). See Black
        Walnut leaves 2 00

485. Water Avens, *Geum rivale* 1 00
    Astringent, tonic. Dose, 15 to 45 minims.

486. Water Eryngo, *Eryngium aquaticum* . . 1 05
    Diaphoretic, expectorant.
    Dose, 20 to 40 minims.

487. Water Fennel seed, *Oenanthe Phellan-
drium* 1 80
    Expectorant, diuretic, mild narcotic.
    Dose, 5 to 10 minims.

488. Water Hemlock, *Cicuta maculata* . . . 1 50
    Sedative, narcotic. Dose, 3 to 5 min

489. Watermelon seed, *Cucumis Citrullus* . . 1 80
    Taenifuge, diuretic, demulcent.
    Dose, 2 fluidrachms to 2 fluidounces, ʒij.

490. Water Pepper, *Polygonum acre* 1 00
    Stimulant, diuretic, emmenagogue.
    Dose, 60 to 120 minims.

491. White Ash, American, *Fraxinus Americana* 1 20
    Emmenagogue. Dose, 15 to 20 minims.

492. White Bryony, *Bryonia alba* 1 50
    Hydragogue cathartic.
    Dose, 10 to 30 minims.

Per lb.

493. White Hellebore, *Veratrum album* . . . $1 20
  Powerful depressant, irritant.
  Dose, 1 to 3 minims.

494. White Oak, *Quercus alba* 1 00
  Tonic, astringent. Dose, 15 to 60 minims.

495. White Pine, *Pinus Strobus* 1 00
  Antiseptic, expectorant.
  Dose, 30 to 60 minims.

496. White Pine Comp., for Syrup 2 00
  100 Cc. represent— 1 fl. oz. represents—
  26.33 Gm. White Pine. . . . 120 grs.
  26.33 Gm. Wild Cherry . . . 120 grs.
  3.51 Gm. Balm of Gilead buds., 16 grs.
  3.51 Gm. Spikenard . . . . . 16 grs.
  3.51 Gm. Blood-root. . . . . 16 grs.
  1.76 Gm. Sassafras . . . . . 8 grs.
  3.12 Cc. Chloroform . . . 15 m.
  0.17 Gm. Morphine acetate. . . 0.8 gr.
  Four times as strong as the Syrup.
  See notes 50, 104.

497. White Pond-lily, *Nymphaea odorata* . . 1 00
  Astringent. Dose, 15 to 30 minims.

498. White Poplar, *Populus tremuloides* . . . 1 00
  Tonic, febrifuge. Dose, 30 to 60 minims.

499. White Vervain, *Verbena urticaefolia* . . 2 50
  Febrifuge. Dose, 20 to 40 minims.

500. White Willow bark, *Salix alba* 1 00
  Tonic, febrifuge, astringent.
  Dose, 15 to 60 minims.

501. Wild Bergamot, *Monarda fistulosa* . . . 2 00
  Antiperiodic, diaphoretic.
  Dose, 15 to 60 minims.

502. Wild Cherry, *Prunus serotina* 1 05
  Tonic, sedative, pectoral.
  Dose, 30 to 60 minims.

503. Wild Cherry, detannated 1 40
  Tonic, sedative, pectoral.
  Dose, 30 to 60 minims.

504. Wild Cherry, for U. S. P. Syrup. . . . 1 05
  Four times as strong as the Syrup.
  See note 50.

Per lb.

505. Wild Cherry Comp. $1 05
      100 Cc. represent— 1 fl. oz. represents—
        52.67 Gm. Wild Cherry . . . 240 grs.
        26.33 Gm. Horehound . . . . 120 grs.
        13.17 Gm. Lettuce. . . . . 60 grs.
        12.28 Gm. Blood-root . . . . 56 grs.
          .88 Gm. Veratrum Viride . . 4 grs.
      Expectorant, tonic, sedative.
      See note 50.   Dose, 30 to 60 minims.

    , Wild Ginger (197).   See Canada Snake-
        root 1 05

506. Wild Indigo, *Baptisia tinctoria* 1 00
      Laxative, emetic, antiseptic.
      Dose, 5 to 15 minims.

507. Wild Lettuce, *Lactuca Canadensis* . . . 1 20
      Mild soporific.   Dose, 20 minims.

508. Wild Yam, *Dioscorea villosa* 1 00
      Diaphoretic, antispasmodic.
      Dose, 15 to 60 minims.

509. Willow-herb, *Epilobium angustifolium* : 1 20
      Astringent.   Dose, 30 to 60 minims.

510. Wintergreen, *Gaultheria procumbens* . . 1 00
      Stimulant, astringent, diuretic.
      Dose, 15 to 60 minims.

511. Witch Hazel, *Hamamelis Virginiana* . 1 00
      For external use.

512. Wood Betony, *Stachys Betonica* . . . 2 50
      Carminative, astringent.
      Dose, 30 to 60 minims.

513. Wormseed, American, *Chenopodium am-
        brosioides*, var. *anthelminticum* . . . 1 00
      Anthelmintic.   Dose, 15 to 60 minims.

    Wormseed, Levant (290).   See Levant
        Wormseed 1 50

    Wormwood (2).   See Absinthium . . . 1 00

    Xanthoxylum (373).   See Prickly Ash
        bark 1 10

514. Yarrow, *Achillea millefolium* 1 00
      Alterative, astringent, tonic.
      Dose, 30 to 60 minims.

515. Yellow Dock, *Rumex crispus* 1 10
      Tonic, alterative, laxative.
      Dose, 15 to 60 minims.

Per lb.

516. Yellow Dock Comp., for Eclec. Disp.
     Syrup $1 50
     100 Cc\ represent— 1 fl. oz. represents—
         52.67 Gm. Yellow Dock . . . . 240 grs.
         26.33 Gm. False Bittersweet . . 120 grs.
         13.17 Gm. American Ivy . . . 60 grs.
         13.17 Gm. Figwort . . . . . . 60 grs.
     Four times as strong as the Syrup.
     See note 50.
517. Yellow Parilla, *Menispermum Canadense.* 1 20
     Tonic, alterative, diuretic.
     Dose, 15 to 60 minims.
518. Yellow Pond Lily, *Nuphar advena* . . . 1 00
     Astringent. Dose, 30 to 60 minims.
519. Yerba Buena, *Micromeria Douglasii* . . 2 50
     Tonic, emmenagogue.
     Dose, 30 to 120 minims.
520. Yerba Reuma, *Frankenia grandifolia* . 2 50
     Astringent. Dose, 10 to 20 minims.
521. Yerba Santa, *Eriodictyon glutinosum* . . 2 50
     Tonic, expectorant. Dose, 30 to 60 minims.
522. Yerba Santa, Aromatic, for Syrup . . . 1 75
     For masking the taste of Quinine.
     See note 105.
523. Yerba Santa, soluble 2 50
     See note 50.
     Zea (175). See Corn Silk 1 50
524. Zedoary, *Curcuma Zedoaria.* 2 00
     Aromatic, stimulant. Dose, 10 to 30 minims.
     Zingiber (226). See Ginger 1 50
525. Lupulin Comp. 1 80

SOLID EXTRACTS.

We invite special attention to our line of solid extracts listed in the following pages. They are prepared from prime drugs by methods especially adapted to their requirements. In the cases of those drugs containing volatile principles or those which may be otherwise injured by heat, vacuum apparatus is used so that concentration may be effected without detriment to the resulting product.

Alkaloidal extracts are made to conform to a fixed standard, thereby assuring their therapeutic activity and uniformity of strength. Our solid extracts are unsurpassed by any now on the market, and have always given entire satisfaction.

No.		Per 1 ℔. Jar.	Per ¼ ℔. Jar.	Per 1 oz. Jar.
1	Aconite leaves, assayed . . Standard: 1:400 dilution, physiological test. See note 107. Dose, 1-4 to 1 gr.	$4 25	$1 15	$ 35
2	Aconite root, assayed . Standard: 1:5000 dilution, physiological test. See note 107. Dose, 1-20 to 1-10 gr.	4 25	1 15	35
3	Aloes . . . . See note 107. Dose, 1 to 5 grs.	2 70	75	25
4	American Hellebore, assayed Standard : 6.% alkaloids, by weight. See note 107. Dose, 1-8 to 1-4 gr.	3 60	1 00	30
5	Angelica. . . . . Dose, 1 to 5 grs.	3 75	1 05	35
6	Arnica flowers . . . . . . Dose, 1 to 5 grs.	3 00	85	25
7	Arnica root . . . . Dose, 1 to 5 grs.	3 60	1 00	30
8	Balmony . . . . Dose, 2 to 5 grs.	3 00	80	30

No.		Per 1 ℔. Jar.	Per ¼ ℔. Jar.	Per 1 oz. Jar.
9	Belladonna leaves, assayed Standard: 1.6% alkaloids, by titration. See note 107. Dose, 1-8 to 1-2 gr.	$3 60	$1 00	$ 30
10	Belladonna root, assayed Standard: 2.5% alkaloids, by titration. See note 107. Dose, 1-16 to 1-4 gr.	4 00	1 10	35
11	Berberis Aquifolium Dose, 2 to 6 grs.	5 00	1 35	40
12	Bitter-root Dose, 1 to 4 grs.	3 35	90	30
13	Bittersweet. Dose, 5 to 15 grs.	2 50	70	25
14	Blackberry-root bark Dose, 3 to 10 grs.	1 80	55	20
15	Black Cohosh Dose, 3 to 10 grs.	3 00	85	25
16	Black Haw Dose, 3 to 10 grs.	3 60	1 00	30
17	Black Hellebore Dose, 1 to 4 grs.	3 00	85	25
18	Bladder-wrack Dose, 5 to 30 grs.	3 00	85	25
19	Blood-root, assayed Standard: 17.5% alkaloids, by titration. See note 107. Dose, 1 to 5 grs.	3 60	1 00	30
20	Blue Cohosh. Dose, 2 to 4 grs.	2 10	60	.20
21	Blue Flag. Dose, 2 to 4 grs.	3 00	85	25
22	Boneset. Dose, 10 to 25 grs.	2 50	70	25
23	Buchu Dose, 3 to 10 grs.	4 80	1 30	40
24	Buckthorn bark Dose, 3 to 10 grs.	3 00	85	25
25	Burdock root Dose, 6 to 20 grs.	2 40	70	25
26	Butternut Dose, 3 to 10 grs.	2 50	70	25

No.		Per 1 ℔. Jar.	Per ¼ ℔. Jar.	Per 1 oz. Jar.
27	Calabar Bean, assayed			$1 90
	Standard: 2.75% physostigmine, by weight. See note 107. Dose, 1-8 to 1-4 gr.			
28	Calendula	5 40	1 45	40
	Dose, 5 to 10 grs.			
29	Canadian Hemp	4 00	1 10	35
	Dose, 1 to 4 grs.			
30	Cannabis Indica	9 60	2 50	70
	Dose, 1-8 to 1 gr.			
31	Cascara Sagrada	5 00	1 35	40
	Dose, 2 to 10 grs.			
32	Cascara Sagrada, tasteless	5 00	1 35	40
	Dose, 2 to 8 grs.			
33	Cascarilla	4 50	1 20	35
	Dose, 5 to 10 grs.			
34	Catechu	2 70	75	25
	Dose, 5 to 10 grs.			
35	Celandine	4 00	1 05	35
	Dose, 5 to 10 grs.			
36	Chamomile, German . . .	3 45	90	30
	Dose, 5 to 10 grs.			
37	Chamomile, Roman . . .	6 00	1 60	45
	Dose, 5 to 10 grs.			
38	Chirata	6 00	1 60	45
	Dose, 5 to 10 grs.			
39	Cinchona, assayed	9 00	2 35	65
	Standard: 20.% total alkaloids, by weight. See note 107. Dose, 5 to 20 grs.			
40	Cinchona, pale, assayed . .	5 00	1 35	40
	Standard: 15.% total alkaloids, by weight. See note 107. Dose, 5 to 20 grs.			
41	Cinchona, red, assayed . .	9 00	2 35	65
	Standard: 17.% total alkaloids, by weight. See note 107. Dose, 5 to 20 grs.			
42	Clover, Red	3 00	85	25
	Dose, 5 to 20 grs.			
43	Coca, assayed	9 00	2 35	65
	Standard: 2.% cocaine, by weight. See note 107. Dose, 5 to 20 grs.			

No.		Per 1 ℔. Jar.	Per ¼ ℔. Jar.	Per 1 oz. Jar.
44	Colchicum root, assayed . . .	$3 60	$1 00	$ 30
	Standard: 2.5% colchicine, by weight. See note 107.			
	Dose, 1-4 to 1 gr.			
45	Colchicum seed, assayed . .	5 50	1 35	45
	Standard: 3.% colchicine, by weight. See note 107.			
	Dose, 1-4 to 1 gr.			
	Colocynth. See Powdered Extracts			. .
	Colocynth Comp. See Powdered Extracts			. .
46	Columbo	3 50	95	30
	Dose, 1 to 4 grs.			
47	Comfrey root	4 50	1 25	45
	Dose, 5 to 10 grs.			
48	Conium fruit, assayed . . .	3 00	85	25
	Standard: 3.% coniine, by titration. See note 107.			
	Dose, 1 to 3 grs.			
49	Conium leaves	2 50	70	25
	Dose, 2 to 5 grs.			
50	Corn-silk	5 25	1 35	40
	Dose, 5 to 20 grs.			
51	Cotton-root bark	4 00	1 10	35
	Dose, 3 to 15 grs.			
52	Cranesbill	3 00	85	25
	Dose, 5 to 10 grs.			
53	Cubeb	7 00	1 85	50
	Dose, 2 to 10 grs.			
54	Culver's-root	4 50	1 20	35
	Dose, 3 to 10 grs.			
55	Damiana	5 40	1 45	40
	Dose, 5 to 15 grs.			
56	Dandelion	1 35	40	15
	Dose, 10 to 30 grs.			
56	Digitalis	2 40	70	25
	Dose, 1-4 to 1-2 gr.			
58	Dogwood	2 75	75	25
	Dose, 2 to 8 grs.			
59	Elder bark	6 00	1 50	50
	Dose, 5 to 10 grs.			

No.		Per 1 ℔. Jar.	Per ¼ ℔. Jar.	Per 1 oz. Jar.
60	Elder flowers . . . . . . Dose, 5 to 10 grs.	$2 25	$ 65	$ 25
61	Elecampane . . . . Dose, 5 to 10 grs.	4 15	1 20	35
62	Ergot . . . . Dose, 1 to 5 grs.	9 00	2 35	65
63	Eucalyptus . . . . Dose, 3 to 10 grs.	3 60	1 00	30
64	Fringe-tree . . . . Dose, 5 to 15 grs.	3 00	85	25
65	Fumitory . . . . Dose, 4 to 10 grs.	5 25	1 35	45
66	Galanga . . . . Dose, 5 to 10 grs.	5 20	1 35	40
67	Galls . . . . Dose, 5 to 10 grs.	5 65	1 45	45
68	Gelsemium, assayed . . . Standard: 3.5% alkaloids, by titration. See note 107. Dose, 1-4 to 2 grs.	3 60	1 00	30
69	Gentian . . . . Dose, 5 to 10 grs.	1 20	40	15
70	Golden Seal, assayed . . . Standard: 2.5% alkaloids, by titration. See note 107. Dose, 3 to 10 grs.	7 30	1 90	55
71	Grindelia . . . . Dose, 5 to 15 grs.	3 60	1 00	30
72	Guaiac wood . . . . Dose, 5 to 10 grs.	6 75	1 75	50
73	Guarana, assayed . . . . . Standard: 20.% caffeine, by weight. See note 107. Dose, 3 to 10 grs.			1 35
74	Henbane, assayed . . . . Standard: .5% alkaloids, by weight. Dose, 1-2 to 2 grs.	3 60	1 00	30
75	Hops . . . . Dose, 3 to 10 grs.	4 80	1 30	40
76	Horehound . . . . Dose, 5 to 15 grs.	2 70	75	25
77	Hydrangea . . . . Dose, 3 to 10 grs.	4 80	1 35	40

No.		Per 1 ℔. Jar.	Per ¼ ℔. Jar.	Per 1 oz. Jar.
78	Ignatia, assayed	$ . .	$ . .	$ 75
	Standard: 20.% alkaloids, by titration. See note 107.			
	Dose, 1-10 to 1-2 gr.			
	Indian Hemp (30)	9 60	2 50	70
	Dose, 1-8 to 1 gr.			
79	Indian Physic	. .	. . .	. 65
	Dose, 2 to 5 grs.			
80	Ipecac, assayed	. .	. .	2 00
	Standard: 10.% emetine, by titration. See note 107.			
	Dose, 1-10 to 1-2 gr.			
81	Jaborandi, assayed	9 00	2 35	65
	Standard: 3.5% pilocarpine, by weight. See note 107.			
	Dose, 3 to 10 grs.			
82	Jalap	4 80	1 30	40
	Dose, 2 to 5 grs.			
83	Jamaica Dogwood	. .	2 85	75
	Dose, 2 to 10 grs.			
84	Juniper berries	2 25	65	25
	Dose, 20 to 30 grs.			
85	Kavakava	6 00	1 60	45
	Dose, 2 to 6 grs.			
86	Kola, assayed	6 00	1 60	45
	Standard: 10.% caffeine, by weight. See note 107.			
	Dose, 5 to 10 grs.			
87	Kousso	. .	. .	1 00
	Dose, 15 to 45 grs.			
88	Life-root	2 50	70	25
	Dose, 5 to 10 grs.			
89	Lobelia	3 00	85	25
	Dose, 1-2 to 2 grs.			
90	Male-fern	3 60	1 00	30
	Dose, 15 to 40 grs.			
91	Mandrake	3 50	95	30
	Dose, 1 to 4 grs.			
92	Musk-root	. .	. .	90
	Dose, 2 to 5 grs.			
93	Nux Vomica, assayed . . .	3 60	1 00	30
	Standard: 15.% alkaloids, by titration. See note 107.			
	Dose, 1-10 to 1 gr.			

No.		Per 1 ℔. Jar.		Per ¼ ℔. Jar.		Per 1 oz. Jar.
94	Opium, assayed	$.	.	$ . . .		$1 20
	Standard: 18.% morphine. See note 107. Dose, 1-4 to 1 gr.					
95	Ox-gall	3	00	85		25
	Dose, 5 to 10 grs.					
96	Pipsissewa	2	40	70		25
	Dose, 3 to 10 grs.					
97	Poke root	3	20	90		30
	Dose, 1 to 5 grs.					
98	Prickly Ash bark	3	60	1	00	30
	Dose, 3 to 10 grs.					
99	Pulsatilla		. .	. .		45
	Dose, 1-4 to 1 gr.					
100	Quassia	4	20	1	15	35
	Dose, 1 to 4 grs.					
101	Quebracho.	9	00	2	50	70
	Dose, 1 to 3 grs.					
102	Queen-of-the-Meadow . . .	3	00	85		25
	Dose, 3 to 10 grs.					
103	Raspberry leaves	2	50	70		25
	Dose, 5 to 10 grs.					
104	Rhatany	2	40	70		25
	Dose, 2 to 4 grs.					
105	Rhubarb	5	00	1	35	40
	Dose, 2 to 10 grs.					
106	Rue	3	00	85		25
	Dose, 2 to 5 grs.					
107	Sarsaparilla	4	50	1	20	35
	Dose, 5 to 10 grs.					
108	Sarsaparilla Comp.	3	00	85		25
	Dose, 5 to 10 grs.					
109	Savine	1	80	55		20
	Dose, 1 to 4 grs.					
110	Scullcap.	4	00	1	10	35
	Dose, 5 to 10 grs.					
111	Senega		. .	. .		45
	Dose, 1 to 3 grs.					
112	Senna.	3	50	95		30
	Dose, 10 to 20 grs.					
113	Sheep-sorrel	3	50	95		30
	Dose, 5 to 10 grs.					
114	Star-grass	5	50	1	45	45
	Dose, 2 to 5 grs.					

No.		Per 1 ℔. Jar.	Per ¼ ℔. Jar.	Per 1 oz. Jar.
115	Stillingia . . . . Dose, 2 to 5 grs.	$3 50	$. 95	$. 30
116	Stramonium leaves, assayed. Standard: 1.5% alkaloids, by weight. See note 107. Dose, 1-4 to 1 gr.	2 40	70	25
117	Stramonium seed, assayed . Standard: 3.5% alkaloids, by weight. See note 107. Dose, 1-8 to 1-2 gr.	4 50	1 20	35
	Sumbul (92). . . . . Dose, 2 to 10 grs.			90
118	Tobacco. . . . . Dose, 1-4 to 1 gr.			45
119	Uva Ursi . . . . Dose, 5 to 15 grs.	2 40	70	25
120	Valerian. . . . . Dose, 5 to 10 grs. 70	3 00	85	25
	Veratrum Viride, assayed (4) Standard: 6.% alkaloids, by weight. See note 107. Dose, 1-8 to 1-4 gr.	3 60	1 00	30
121	Wahoo . . . . Dose, 5 to 15 grs.	3 60	1 00	30
122	White Hellebore . . . . . Dose, 1-8 to 1-4 gr.	3 60	1 00	30
123	Witch Hazel. . . . . . . Dose, 5 to 15 grs.	2 40	70	25
124	Wormwood . . . . Dose, 1 to 10 grs.	2 75	75	25
125	Yellow Dock. . . . . . . Dose, 5 to 20 grs.	2 40	70	25
126	Yellow Parilla . . . . . . Dose, 5 to 10 grs.			40
127	Yerba Santa . . . . Dose, 5 to 10 grs.	3 60	1 00	30

POWDERED EXTRACTS.

Powdered extracts possess certain advantages over those of pilular consistency, the greatest of which is the facility with which they may be dispensed. At the same time, as they are somewhat hygroscopic, it is necessary to keep the containers tightly stoppered in order to preserve them in the powdered condition.

Our line of these goods includes all that have been found meritorious, and may be employed with the certainty that none are better. Those containing alkaloids are adjusted to a fixed standard, thereby insuring uniformity of strength.

No.		Per oz. vial.
1.	Aconite leaves, assayed . . . . $	40
	Standard: 1:400 dilution, physiological test. See note 92. Dose, 1-4 to 1 gr.	
2.	Aconite root, assayed	40
	Standard: 1:5000 dilution, physiological test. See note 92. Dose, 1-20 to 1-10 gr.	
3.	Aloes.	40
	Dose, 1 to 5 grs.	
4.	Belladonna leaves, assayed	45
	Standard: 1.6% alkaloids, by titration. See note 92. Dose, 1-8 to 1-2 gr.	
5.	Belladonna root, assayed	50
	Standard: 2.5% alkaloids, by titration. See note 92. Dose, 1-16 to 1-4 gr.	
6.	Berberis Aquifolium	60
	Dose, 2 to 6 grs.	
7.	Blackberry-root bark.	45
	Dose, 3 to 10 grs.	
8.	Black Cohosh	40
	Dose, 3 to 10 grs.	
9.	Black Haw	50
	Dose, 3 to 10 grs.	
10.	Blood-root, assayed	40
	Standard: 1.75% alkaloids, by titration. See note 92. Dose, 1 to 5 grs.	
11.	Cannabis Indica	75
	See note 63. Dose, 1-2 to 2 grs.	

POWDERED EXTRACT.

No.		Per oz. vial.

12. Cáscara Sagrada 60
    Dose, 2 to 10 grs.

13. Cinchona, assayed 75
    Standard: 20.% alkaloids, by weight.
    See note 92.   Dose, 5 to 20 grs.

14. Cinchona, red, assayed 75
    Standard: 17.% alkaloids, by weight.
    See note 92.   Dose, 5 to 20 grs.

15. Coca, assayed 75
    Standard: 2.% cocaine, by weight.
    See note 92.   Dose, 5 to 20 grs.

16. Colchicum root, assayed 45
    Standard: 2.5% colchicine, by weight.
    See note 92.   Dose, 1-4 to 1 gr.

17. Colchicum seed, assayed 70
    Standard: 3.% colchicine, by weight.
    See note 92.   Dose, 1-4 to 1 gr.

18. Colocynth 60
    Dose, 1 to 2 grs.

19. Colocynth Comp. 30
    Dose, 5 to 20 grs.

20. Columbo 50
    Dose, 1 to 4 grs.

21. Conium fruit, assayed 40
    Standard: 3.% coniine, by titration.
    See note 92.   Dose, 1 to 3 grs.

22. Conium leaves 40
    Dose, 2 to 5 grs.

23. Culver's root 45
    Dose, 3 to 10 grs.

24. Dandelion 40
    Dose, 10 to 30 grs.

25. Digitalis 40
    Dose, 1-4 to 1-2 gr.

26. Ergot 75
    Dose, 1 to 5 grs.

27. Eucalyptus 50
    Dose, 3 to 10 grs.

28. Gelsemium, assayed 50
    Standard: 3.5% alkaloids, by titration.
    See note 92.   Dose, 1-4 to 2 grs.

29. Gentian 40
    Dose, 5 to 10 grs.

SUGAR-COATED AND GELATIN-COATED.

No.		Per oz. vial.

30. Golden Seal, assayed . . . . . $ 65
   Standard: 12.5% alkaloids, by titration.
   See note 92. Dose, 3 to 10 grs.

31. Henbane, assayed. 45
   Standard: .5% alkaloids, by weight.
   See note 92. Dose, 1-2 to 2 grs.
   Indian Hemp (11), See Cannabis Indica. 75

32. Ipecac, assayed 2 25
   Standard: 10.% emetine, by titration.
   See note 92. Dose, 1-10 to 1-2 gr.

33. Lobelia 40
   Dose, 1-2 to 2 grs.

34. Mandrake 40
   Dose, 1 to 4 grs.

35. Nux Vomica, assayed 40
   Standard: 15.% alkaloids, by titration.
   See note 92. Dose, 1-10 to 1 gr.

36. Opium, assayed 1 25
   Standard: 18.% morphine.
   See note 92. Dose, 1-4 to 1 gr.

37. Prickly Ash 50
   Dose, 3 to 10 grs.

38. Quassia 50
   Dose, 1 to 4 grs.

39. Rhatany 40
   Dose, 2 to 4 grs.

40. Rhubarb 45
   Dose, 2 to 10 grs.

41. Sarsaparilla 65
   Dose, 5 to 10 grs.

42. Senega 50
   Dose, 1 to 3 grs.

43. Senna 50
   Dose, 10 to 20 grs.

44. Stillingia 50
   Dose, 2 to 5 grs.

45. Stramonium leaves, assayed . . 40
   Standard: 1.5% alkaloids, by weight.
   See note 92. Dose, 1-4 to 1 gr.

46. Stramonium seed, assayed . . . 40
   Standard: 3.5% alkaloids, by weight.
   See note 92. Dose, 1-8 to 1-2 gr.

47. Valerian 40
   Dose, 5 to 10 grs.

SUGAR-COATED AND GELATIN-COATED PILLS.

Next to the activity of the ingredients, the prime requisite of pills is ready solubility. There is no more crucial test to which they may be subjected. Common sense teaches that the friability of a pill has but little if anything more to do with its solubility than its color has. An exhaustive series of experiments recently performed in our laboratory shows conclusively that mass pills when properly made satisfy all the requirements as to solubility, being far superior to friable pills in that respect. For further information see note 88.

The solubility of our pills has never been questioned. Nearly half a century of pill-making has given us a more extended experience than most other manufacturers possess, and we have endeavored to use this to good advantage. Consequently, it was no surprise to us when in the rigid tests already mentioned (over three thousand in number) the solubility of our pills was shown to be absolutely unsurpassed.

In deference to the general custom, we have for many years made all sugar-coated pills round and gelatin-coated pills oval. Believing, however, that the oval is the more acceptable shape, we intend to make all our pills in that form hereafter, both sugar-coated and gelatin-coated. At the same time we reserve the right to supply either round or oval sugar-coated pills on orders which do not specify the shape desired.

We will be pleased to quote special figures on pills made up from private formula, in lots of not less than three thousand. Our pills are made by machinery of the most approved design and cannot be excelled for elegant appearance and accuracy of dose, while our name is sufficient guaranty of the purity and high quality of the ingredients used in their preparation.

No.		Per 100.	Per 500.
1.	Acetanilid, 5 grs.	$ 45	$2 10

    Antipyretic, analgesic. Dose, 1 to 2.

| 2. | Acetanilid Comp. | 45 | 2 10 |

    Acetanilid . . . . . . 2 grs,
    Monobromated Camphor. . 1-2 gr.
    Citrated Caffeine . . . 1-2 gr.
    Sedative, analgesic. Dose, 1 to 2.

Acid, Arsenous. See Arsenous acid.

Acid, Carbolic. See Carbolic acid.

| 3. | Aconite root, Extract, 1-4 gr. | 20 | 85 |

    Antipyretic, sedative. Dose, 1 to 2.

| 4. | Aconitine, cryst., 1-500 gr. | 25 | 1 10 |

    Antipyretic, sedative. Dose, 1 to 4.

| 5. | Aconitine, cryst., 1-200 gr. | 25 | 1 10 |

    Dose, 1 to 2.

| 6. | Agaricin, 1-6 gr. | 35 | 1 60 |

    Anhydrotic. Dose, 1 to 3.

Aiken's Copaiba. See Copaiba

| | Comp., " A " (245). | 55 | 2 60 |

Aiken's Tonic. See Tonic, "A,"

| | Aiken's (641) : | 55 | 2 60 |

| 7. | Aloes, 1 gr. | 20 | 85 |

    Purgative. Dose, 1 to 5.

| 8. | Aloes, U. S. P. | 25 | 1 10 |

    .13 Gm. Purified Aloes . . . 2 grs.
    .13 Gm. Powd. Soap . . . . 2 grs.
    Purgative. Dose, 1 to 3.

| 9. | Aloes, 4 grs. | 25 | 1 10 |

    Dose, 1 to 2.

| 10. | Aloes and Asafetida, U. S. P. | 25 | 1 10 |

    .09 Gm. Purified Aloes . . . 1 2-5 grs.
    .09 Gm. Asafetida . . . . 1 2-5 grs.
    .09 Gm. Powd. Soap. . . . 1 3-5 grs.
    Antispasmodic, purgative.
    Dose, 1 to 3.

| 11. | Aloes and Asafetida, 4 grs., B. P. | 20 | 85 |

    Sugar-coated only. See note 99.
    Powdered Socotrine Aloes . . . 1 gr.
    Asafetida . . . . 1 gr.
    Powdered Hard Soap . . . . 1 gr.
    Confection Roses . . . . . 1 gr.
    Antispasmodic, purgative.
    Dose, 1 to 3.

No. 1	Per 100.	Per 500.

12. Aloes and Asafetida, 5 grs., B.P. . $ 35 . $1.60
    Gelatin-coated only. See note 89.
    Powdered Socotrine Aloes . 1 1-4 grs.
    Asafetida . . . . 1 1-4 grs.
    Powdered Hard Soap . . . 1 1-4 grs.
    Confection Roses . . . . 1 1-4 grs.
    Antispasmodic, purgative.
    Dose, 1 to 3.

13. Aloes, Barbados, B. P. 30   1 35
    Sugar-coated only. See note 90.
    Powdered Barbados Aloes . . 2 grs.
    Oil Caraway . . . . 1-8 gr.
    Powdered Hard Soap . . . 1 gr.
    Confection Roses . . . . . 1 gr.
    Purgative. Dose, 1 to 2.

14. Aloes and Iron, U. S. P. . . . 25   1 10
    .07 Gm. Dried Ferrous sulph. 1 1-12 grs.
    .07 Gm. Purified Aloes . . . 1 1-12 grs.
    .07 Gm. Aromatic Powder . 1 1-12 grs.
    Confection of Rose . . . . . q. s.
    Tonic, purgative. Dose, 1 to 3.

15. Aloes and Iron, B. P. 30   1 35
    Sugar-coated only. See note 90.
    Iron sulphate . . . . 1-2 gr.
    Powdered Barbados Aloes . . 2-3 gr.
    Comp. Cinnamon Powder . . . 1 gr.
    Confection Roses . . . . . 1-3 gr.
    Tonic, purgative. Dose, 1 to 4.

16. Aloes and Iron, 5 grs., B. P. . . 50   2 35
    Gelatin-coated only. See note 89.
    Powdered Barbados Aloes . 20-21 gr.
    Iron sulphate . . . . . . 5-7 gr.
    Comp. Cinnamon Powder . 1 3-7 grs.
    Confection Roses . . . . 1 19-21 grs.
    Tonic, purgative. Dose, 1 to 3.

17. Aloes and Mastic, U. S. P. . . . 30   1 35
    .13 Gm. Purified Aloes . . . . 2 grs.
    .04 Gm. Mastic . . . . . . 3-5 gr.
    .03 Gm. Red Rose . . . . . 1-2 gr.
    Stimulating cathartic. Dose, 1 to 2.

18. Aloes and Myrrh, U. S. P. . . . 25   1 10
    .13 Gm. Purified Aloes . . . . 2 grs.
    .06 Gm. Myrrh . . . . 1 gr.
    .04 Gm. Aromatic Powder . . 3-5 gr.
    Emmenagogue, cathartic.
    Dose, 1 to 2.

No.		Per 100.	Per 500.

19. Aloes and Myrrh, B. P. . . . . . . $ 25 . $1. 10
    Sugar-coated only. See note 90.
    Socotrine Aloes . . . . 2 grs.
    Saffron . . . . 1-2 gr.
    Myrrh . . . . 1. 1 gr.
    Emmenagogue, cathartic.
    Dose, 1 to 3.

20. Aloes and Myrrh, 5 grs., B. P. . . . 40 1 85
    Gelatin-coated only. See note 89.
    Socotrine Aloes . . . . . . 2 6-7 grs.
    Myrrh . . . . 1 3-7 grs.
    Saffron . . . . 5-7 gr.
    Emmenagogue, cathartic.
    Dose, 1 to 2.

21. Aloes and Nux Vomica 30 1 35
    Socotrine Aloes . . . . . 1 1-2 grs.
    Extract Nux Vomica . . . 1-2 gr.
    Tonic, laxative. Dose, 1 to 2.

22. Aloes, Nux Vomica and Bella-
    donna. 35 1 60
    Socotrine Aloes . . . . . 1 1-2 grs.
    Extract Nux Vomica . . . 1-2 gr.
    Extract Belladonna . . . . 1-8 gr.
    Tonic, laxative. Dose, 1 to 3.

23. Aloes and Rhubarb 30 1 35
    Socotrine Aloes . . . . . . 1 gr.
    Extract Rhubarb . . . . . 1 gr.
    Soap . . . . 1 gr.
    Oil Chamomile . . . . . 1-3 gr.
    Cathartic. Dose, 1 to 3.

24. Aloes, Socotrine, B. P. 25 1 00
    Sugar-coated only. See note 90.
    Powdered Socotrine Aloes . . 2 grs.
    Oil Nutmeg . . . . 1-8 gr.
    Powdered Hard Soap . . . . 1 gr.
    Confection Roses . . . . . . 1 gr.
    Purgative. Dose, 1 to 3.

25. Aloin, 1-10 gr. 20 85
    Cathartic. Dose, 1 to 3.

26. Aloin, 1-5 gr. 25 1 10
    Dose, 1 to 2.

27. Aloin, 1-4 gr. 25 1 10
    Dose, 1 to 2.

28. Aloin, 1-2 gr. 35 1 60
    Dose, 1 to 2.

29. Aloin, 1 gr. 60 2 85
    Dose, 1 to 2.

PILES.

No.		Per 100.	Per 500.
30.	Aloin Comp. . . . . $	30	$1 35

    Aloin . . . . 1-8 gr.
    Resin Podophyllum . . . . 1-8 gr.
    Ext. Belladonna . . . . . . 1-4 gr.
    Laxative. Dose, 1 to 3.

| 31. | Aloin Comp. with Strychnine . . . | 40 | 1 85 |

    Aloin . . . . 1-8 gr.
    Resin Podophyllum . . . . 1-8 gr.
    Ext. Belladonna . . . . . . 1-8 gr.
    Strychnine . . . . 1-80 gr.
    Oleoresin Capsicum . . . . . 1-10 gr.
    Laxative, tonic. Dose, 1 to 3.

| 32. | Aloin and Strychnine . . . . | 40 | 1 85 |

    Aloin . . . . 1-5 gr.
    Strychnine . . . . 1-60 gr.
    Laxative, tonic. Dose, 1 to 2.

| 33. | Aloin, Strychnine and Belladonna "A" . . . . | 40 | 1 85 |

    Aloin . . . . 1-5 gr.
    Strychnine . . . . 1-60 gr.
    Ext. Belladonna . . . . . . 1-8 gr.
    Laxative, tonic. Dose, 1 to 2.

| 34. | Aloin, Strychnine and Belladonna "B" . . . . | 40 | 1 85 |

    Aloin . . . . 1-10 gr.
    Strychnine . . . . 1-50 gr.
    Ext. Belladonna . . . . . . 1-6 gr.
    Laxative, tonic. Dose, 1 to 2.

| 35. | Aloin, Strychnine and Belladonna "C" . . . . | 40 | 1 85 |

    Aloin . . . . 1-5 gr.
    Strychnine . . . . 1-120 gr.
    Ext. Belladonna . . . . . . 1-8 gr.
    Laxative, tonic. Dose, 1 to 2.

| 36. | Aloin, Strychnine, Belladonna and Cascara Sagrada . . . . . | 50 | 2 35 |

    Aloin . . . . 1-5 gr.
    Strychnine . . . . 1-60 gr.
    Ext. Belladonna . . . . . . 1-8 gr.
    Ext. Cascara Sagrada . . . . 1-2 gr.
    Laxative, tonic. Dose, 1 to 2.

| | Aloin, Strychnine, Belladonna and Ipecac. See Laxatonic (387). | 35 | 1 60 |

No.   Per 100.   Per 500.

37. Alterative . . . . $ 25 $1.10
    Mercury mass . . . . 1 gr.
    Opium . . . . 1-8 gr.
    Ipecac . . . . 1-8 gr.
    Mercurial alterative.   Dose, 1 to 2.

38. Alterative Comp.   50   2 35
    Ext. Smilax Sarsaparilla . . . 2-3 gr.
    Ext. Stillingia . . . . 2-3 gr.
    Ext. Lappa . . . . 2-3 gr.
    Ext. Phytolacca . . . . 2-3 gr.
    Ext. Xanthoxylum . . . . 1-3 gr.
    Tonic, alterative.   Dose, 1 to 3.

    Alvord's.   See Dysmenorrhea.
    (281)   75   3 60

39. Amenorrhea   55   2 60
    Socotrine Aloes . . . . 1 gr.
    Dried Ferrous sulphate . . . 1 gr.
    Ext. Cimicifuga . . . . 1 gr.
    Ext. Cotton Root . . . . 1 gr.
    Emmenagogue.   Dose, 1 to 2.

40. Ammonium bromide, 1 gr.   45   2 10
    Sugar-coated only. See note 90.
    Nervous sedative, hypnotic.
    Dose, 1 to 5.

41. Ammonium chloride, 3 grs.   30   1 35
    Sugar-coated only. See note 90.
    Stimulant, expectorant. . . . . .
    Dose, 1 to 2.

42. Ammonium chloride Comp. . . . 1 25   6 10
    Ammonium chloride. . . . . . . 1 gr.
    Opium . . . . 1-32 gr.
    Benzoic acid . . . . 1-32 gr.
    Ext. Licorice . . . . 3-4 grs.
    Acacia . . . . 1-8 gr.
    Camphor. . . . . 1-50 gr.
    Tartar Emetic . . . . 1-60 gr.
    Oil Anise . . . . 1-32 gr.
    Cough sedative, expectorant.

43. Ammonium picrate, 1-8 gr. 25   1 10
    Antiperiodic, antimalarial.
    Dose, 1 to 3.

44. Ammonium picrate, 1-4 gr. 30   1 35
    Dose, 1 to 3.

45. Ammonium picrate, 1-2 gr. 35   1 60
    Dose, 1 to 3.

46. Ammonium picrate, 1 gr. 45   2 10
    Dose, 1 to 2.

PILLS.

No.		Per 100.	Per 500.
47.	Ammonium valerianate, 1 gr. . . .	$ 40	$1 85

Sugar-coated only. See note 90.
Sedative, antispasmodic.
Dose, 1 to 2.

| 48. | Anderson's Scots . . . . | 25 | 1 10 |

Barbados Aloes . . . . 1 3-5 grs.
Colocynth . . . . 4-15 gr.
Gamboge . . . . 1-15 gr.
Soap. . . . . 1-15 gr.
Oil Anise . . . . 1-30 gr.
Purgative. Dose, 1 to 5.

	Andrews' Anodyne. See Capsicum Comp. (170)	40	1 85
	Andrews' Anti-Grippe. See Anti-Catarrhal (53)	65	3 10
	Andrews' Ox-gall. See Ox-gall Comp. (457).	55	2 60
	Andrews' Tonic Hematic. See Tonic Hematic (643)	75	3 60
49.	Anodyne	45	2 10

Camphor . . . . 1 gr.
Extract Hyoscyamus . . . . . . 1 gr.
Morphine acetate . . . . 1-20 gr.
Oleresin Capsicum . . . . . . 1-20 gr.
Dose, 1 to 2.

| | Anthelmintic. See Santonin and Calomel (604) . . . . | 55 | 2 60 |
| 50. | Anti-Bilious, "A," Barclay's . . | 45 | 2 10 |

Extract Colocynth . . . . . . 1 gr.
Resin Jalap . . . . 1-2 gr.
Guaiac . . . . 1 1-2 grs.
Soap . . . . 3-4 gr.
Oil Caraway, Oil Juniper, aa. . . q. s.
Dose, 1 to 3.

| 51. | Anti-Bilious, "B," Cook's . . . | 25 | 1 10 |

Aloes . . . . 1 gr.
Rhubarb . . . . 1 gr.
Mild Mercurous chloride . . . 1-2 gr.
Soap . . . . 1-2 gr.
Dose, 1 to 3.

| 52. | Anti-Bilious, "C" . . . . . . | 40 | 1 85 |

Comp. Ext. Colocynth . . . 2 1-2 grs.
Resin Podophyllum . . . . . 1-4 gr.
Dose, 1 to 2.

No.		Per 100.	Per 500.

53. Anti-Catarrhal, Andrews'. . . . . $ 65 . $3 10
    Quinine salicylate . . . . 1 gr.
    Arsenous acid . . . . 1-125 gr.
    Extract Belladonna . . . . . 1-33 gr.
    Analgesic, anti-catarrhal, antipy- . . . .
    retic. Dose, 1 to 3.

    Anti-Chill, Watson's. See Chin-
    oidine Comp., "B" (194) . . . 40 . 1 85

54. Anti-Chlorotic . . . . 35 . 1 60
    Potassium chlorate . . . . . . 1-2 gr.
    Mandrake . . . . 1-2 gr.
    Iron chloride . . . . 1-2 gr.
    Myrrh . . . . 1-2 gr.
    Hematic tonic. Dose, 1 to 3.

55. Anti-Constipation, "A," Goss' . . 40 . 1 85
    Resin Podophyllum . . . . . 1-4 gr.
    Extract Butternut . . . . . 1-2 gr.
    Extract Hyoscyamus . . . . 1-8 gr.
    Extract Cascara Sagrada . . . 1-4 gr.
    Extract Colocynth . . . . . 1-4 gr.
    Extract Nux Vomica . . . . . 1-8 gr.
    Extract Gentian . . . . . . 1-4 gr.
    Canadian Hemp . . . . . . 1-2 gr.
    Dose, 1 to 2.

56. Anti-Constipation, "B," Palmer's . 35 . 1 60
    Socotrine Aloes . . . . 1 gr.
    Extract Hyoscyamus . . . . . . 1 gr.
    Extract Nux Vomica . . . . . 1-3 gr.
    Ipecac . . . . 1-10 gr.
    Dose, 1 to 2.

57. Anti-Constipation, "C," Brundage 40 . 1 85
    Extract Nux Vomica . . . . . 1-4 gr.
    Extract Hyoscyamus . . . . . 1-4 gr.
    Extract Belladonna . . . . . 1-10 gr.
    Capsicum . . . . 1-4 gr.
    Resin Podophyllum . . . . . 1-10 gr.
    Dose, 1 to 2.

58. Anti-Dyspeptic, "A" . . . . . . 35 . 1 60
    Strychnine . . . . 1-40 gr.
    Extract Belladonna . . . . 1-10 gr.
    Ipecac . . . . 1-10 gr.
    Mercury mass . . . . 2 grs.
    Compound Extract Colocynth . 2 grs.
    Dose, 1 to 2.

No.	Per 100.	Per 500.

59. Anti-Dyspeptic, "B" . . . . . . $.45, $2 10
   Extract Ignatia . . . . 1-4 gr.
   Extract Cinchona . . . . . 1-4 gr.
   Extract Rhubarb. . . . . . . 1-4 gr.
   Capsicum . . . . 1-4 gr.
   Dose, 1 to 2.

60. Anti-Epileptic . . . . 1 20 . 5 85
   Quinine valerianate . . . . . 1 gr.
   Extract Valerian . . . . 1 gr.
   Zinc valerianate . . . . . . 1-2 gr.
   Iron ferrocyanide. . . . . . 1-2 gr.
   Dose, 1 to 2.

61. Anti-Erotic 1 20   5 85
   Camphor . . . . 1-3 grs.
   Lactucarium . . . . . . 1-4 gr.
   Guarana . . . . 1-2 gr.
   Extract Belladonna . . . . 1-4 gr.
   Scale Opium . . . . 1-4 gr.
   Dose, 1 to 2.

Antifebrin. See Acetanilid.

Anti-Grippe. See Anti-Catarrhal,
  (53) 65   3 10

62. Anti-Lacteous, Dr. Stork's . . . .45 . 2 10
   Sodium acetate . . . . . . 3 grs.
   Camphor . . . . 1 gr.
   Potassium nitrate . . . . . 1 gr.
   Elderberry confection, q. s.
   Antigalactagogue. Dose, 1 to 2.

63. Anti-Malarial, "A" . . . . . . 60 . 2 85
   Quinine sulphate . . . . . 1-2 gr.
   Cinchonine sulphate . . . . 1-2 gr.
   Cinchonidine sulphate . . . 1-2 gr.
   Quinidine sulphate . . . . . 1-2 gr.
   Dose, 1 to 2.

64. Anti-Malarial, "B," McCaw's . 65   3 10
   Quinine sulphate . . . . . 1 gr.
   Dried Ferrous sulphate . . . 1-4 gr.
   Gelsemin (resinoid) . . . . 1-4 gr.
   Arsenous acid . . . . . . 1-80 gr.
   Resin Podophyllum . . . . 1-8 gr.
   Oleoresin Black Pepper . . 1-16 gr.
   Dose, 1 to 2.

No.		Per 100.	Per 500.

65. Anti-Malarial, "C," Maddin's,
milder . . . . $ 50 $2 35
  Gelatin-coated only. See note 89.
  Strychnine . . . . 1-40 gr.
  Reduced Iron . . . . 1 gr.
  Purified Aloes . . . . 1-6 gr.
  Arsenous acid . . . . 1-24 gr.
  Quinine sulphate . . . . . . 1 gr.
  Dose, 1 to 2.

66. Anti-Malarial, "D," Maddin's,
milder, without Aloes 50 2 35
  Gelatin-coated only. See note 89.
  Strychnine . . . . 1-40 gr.
  Reduced Iron . . . . 1 gr.
  Arsenous acid . . . . 1-24 gr.
  Quinine sulphate . . . . . . 1 gr.
  Dose, 1 to 2.

67. Anti-Malarial, "E," Maddin's,
stronger 55 2 60
  Gelatin-coated only. See note 89.
  Strychnine . . . . 1-33 gr.
  Reduced Iron . . . . 1 1-5 grs.
  Purified Aloes . . . . 1-5 gr.
  Arsenous acid . . . . 1-20 gr.
  Quinine sulphate. . . . . . 1 1-5 grs.
  Dose, 1 to 2.

68. Anti-Malarial, "F," Maddin's,
stronger, without Aloes 55 2 60
  Gelatin-coated only. See note 89.
  Strychnine . . . . 1-33 gr.
  Reduced Iron . . . . 1 1-5 grs.
  Arsenous acid . . . . 1-20 gr.
  Quinine sulphate. . . . . . 1 1-5 grs.
  Dose, 1 to 2.

69. Antimony Comp., U. S. P., Plum-
mer's 30 1 35
  .04 Gm. Sulphurated Antimony 3-5 gr.
  .04 Gm. Mild Mercurous chlor. 3-5 gr.
  .08 Gm. Powdered Guaiac . 1 1-5 grs.
  Alterative. Dose, 1 to 3.

70. Antimony and Potass. tart., 1-16 gr. 20 85
  Diaphoretic, expectorant.
  Dose, 1 to 2.

71. Antimony and Potass. tart., 1-10 gr. 20 85
  Dose, 1 to 2.

72. Antimony and Potass. tart., 1-8 gr. 20 85
  Dose, 1 to 2.

No.		Per 100.	Per 500.
73.	Antimony, Sulphurated, 1-4 gr. .$1 45	$2 10	
	Diaphoretic. Dose, 1 to 2.	blim	
74.	Antiperiodic, "A," Stearns' . . . 75	3 60	
	Quinine sulphate . . . . . 3-5 gr.		
	Quinidine sulphate . . . . . 1-5 gr.		
	Cinchonine sulphate . . . . 1-2 gr.		
	Cinchonidine sulphate . . . 1-2 gr.		
	Chinoidine . . . . 1 1-5 grs.		
	Dose, 1 to 3.		
75.	Antiperiodic, "B," Improved. . 40	1 85	
	Cinchonidine sulphate . . . . . 1 gr.		
	Resin Podophyllum . . . . 1-20 gr.		
	Dried Ferrous sulphate . . . 1-2 gr.		
	Strychnine sulphate . . . 1-33 gr.		
	Gelsemin (resinoid) . . . . 1-20 gr.		
	Oleoresin Capsicum . . . 1-10 gr.		
	Dose, 1 to 2.		
76.	Antiperiodic, "C" 60	2 85	
	Myrrh . . . . 2 2-3 grs.		
	Saffron . . . . 1 1-3 grs.		
	Zedoary . . . . 1-3 gr.		
	Quinine sulphate . . . . . 2-3 gr.		
	Camphor . . . . 1-3 gr.		
	Dose, 1 to 3.		
	Antiperiodic, Warburg's. See		
	Warburg's Tincture.		
77.	Antipyrine, 1 gr. 1 20	5 85	
	Antipyretic, analgesic.		
	Dose, 1 to 4.		
78.	Antipyrine, 2 1-2 grs. 3 00	14 85	
	Dose, 1 to 4.		
79.	Antipyrine, 5 grs. 4 50	22 35	
	Dose, 1 to 3.		
80.	Antispasmodic, Warner's . . . 45	2 10	
	Extract Hyoscyamus . . . . 1-2 gr.		
	Monobrom. Camphor . . . . 1-2 gr.		
	Capsicum . . . . 1-2 gr.		
	Morphine acetate . . . . 1-10 gr.		
	Dose, 1 to 2.		
81.	Aperient "A," Drysdale's . . . 35	1 60	
	Rhubarb . . . . 1 1-5 grs.		
	Purified Aloes . . . . . . 1 2-5 grs.		
	Ipecac . . . . 2-5 gr.		
	Nux Vomica . . . . 1-2 gr.		
	Dose, 1 to 2.		

No.   Per 100. Per 500.

82. Aperient "B" . . . . $;50 r. '$2 35
    Extract Nux Vomica . l . . 1-3 gr.
    Extract Hyoscyamus . . . . . 1-2 gr.
    Comp. Extract Colocynth . . 2 grs.
    Dose, 1 to 2.

83. Aperient "C," Bauer's   45   2 10
    Extract Hyoscyamus . . . . 1-3 gr.
    Extract Aloes . . . . 1 gr.
    Comp. Extract Colocynth . . . 1 gr.
    Rochelle Salt . . . . 1 1-2 grs.
    Dose, 1 to 2.

    Aphrodisiac. See Damiana Comp.
    (266) . . . .   90'   4 35

84. Aphrodisiac Improved b.75   3 60
    Phosphorus . . . . 1-50 gr.
    Extract Nux Vomica . . . . 1-8 gr.
    Extract Damiana . . . . 1 gr.
    Extract Gentian . . . . 1 gr.
    Cantharides . . . . 1-2 gr.
    Dose, 1 to 2.

85. Arsenic, Fr. Codex   20   85
    Arsenous acid . . . . . . 5-64 gr.
    Black Pepper . . . . 5-6 gr.
    Alterative, tonic. Dose, 1 to 2.

86. Arsenic sulphide, 1-100 gr. . .   20   85
    Alterative, tonic. Dose, 1 to 2.

87. Arsenic sulphide, 1-60 gr. . . .   20   85
    Dose, 1 to 2.

88. Arsenous acid, 1-100 gr. . . .   20   85
    Antiperiodic, alterative.
    Dose, 1 to 2.

89. Arsenous acid, 1-60 gr.   20   85
    Dose, 1 to 2.

90. Arsenous acid, 1-50 gr.   20   85
    Dose, 1 to 2.

91. Arsenous acid, 1-40 gr.   20   85
    Dose, 1 to 2.

92. Arsenous acid, 1-32 gr.   20   85
    Dose, 1 to 2.

93. Arsenous acid, 1-30 gr.   20   85
    Dose, 1 to 2.

94. Arsenous acid, 1-20 gr.   20   85
    Dose, 1 to 2.

95. Arsenous acid, 1-12 gr.   20   85
    Dose, 1.

No.		Per 100.	Per 500.
96.	Asafetida, 2 grs. . . . . $.25		$1 10
	Antispasmodic, sedative, emmena-		
	gogue. Dose, 1 to 5.		
97.	Asafetida, U. S. P. 25		1 10
	.20 Gm. Asafetida . . . . . 3 grs.		
	.06 Gm. Powdered Soap . . 1 gr.		
	Dose, 1 to 3.		
98.	Asafetida, 4 grs. 25		1 10
	Dose, 1 to 2.		
99.	Asafetida, 5 grs. 35		1 60
	Gelatin-coated only. See note 89.		
	Dose, 1 to 2.		
100.	Asafetida Comp. 45		2 10
	Asafetida . . . . 4-5 gr.		
	Opium . . . . 4-5 gr.		
	Ammonium carbonate . . . . 4-5 gr.		
	Antispasmodic, sedative, expector-		
	ant. Dose, 1.		
101.	Asafetida Comp., 3 grs., B. P. . 30		1 35
	Sugar-coated only. See note 90.		
	Asafetida . . . . 1 gr.		
	Myrrh . . . . 1 gr.		
	Galbanum . . . . 1 gr.		
	Dose, 1 to 2.		
102.	Asafetida Comp., 5 grs., B. P. . 50		2 35
	Gelatin-coated only. See note 89.		
	Asafetida . . . . 1 2-3 grs.		
	Myrrh . . . . 1 2-3 grs.		
	Galbanum . . . . 1 2-3 grs.		
	Dose, 1 to 2.		
103.	Asafetida and Iron 25		1 10
	Asafetida . . . . 2 grs.		
	Dried Ferrous sulphate . . . 1 gr.		
	Antispasmodic, tonic. Dose, 1 to 2.		
104.	Asafetida and Nux Vomica . . 35		1 60
	Asafetida . . . . 3 grs.		
	Extract Nux Vomica . . . 1-4 gr.		
	Antispasmodic, nerve tonic.		
	Dose, 1 to 2.		
105.	Asafetida and Rhubarb 40		1 85
	Asafetida . . . . 1 gr.		
	Rhubarb . . . . 1 gr.		
	Reduced Iron . . . . 1 gr.		
	Sedative, tonic. Dose, 1 to 2.		

No.   Per 100.   Per 500.

106. Asiatic 35 $1 60
  0.0075 Gm. Arsenous acid . . . 1-9 gr.
  0.06 Gm. Black Pepper . . . 1 gr.
  0.015 Gm. Acacia . . . . . 1-4 gr.
  0.02 Gm. Althaea . . . . 3-10 gr.
  Antiperiodic, antimalarial.
  Dose, 1.

107. Astringent, Warner's 35   1 60
  Extract Geranium . . . . . 2 grs.
  Opium . . . . 1-3 gr.
  Oleoresin Ginger. . . . . 1-20 m.
  Oil Peppermint . . . . . 1-20 m.
  Narcotic, sedative. Dose, 1 to 2.

108. Atropine, 1-120 gr. 30   1 35
  Narcotic, sedative. Dose, 1 to 2.

109. Atropine, 1-100 gr. 30   1 35
  Dose, 1 to 2.

110. Atropine, 1-60 gr. 40   1 85
  Dose, 1.

111. Atropine and Morphine sulph.
  "A" 50   2 35
  Atropine . . . . 1-300 gr.
  Morphine sulphate . . . . 1-24 gr.
  Narcotic, anodyne. Dose, 1 to 3.

112. Atropine and Morphine sulph.
  "B" 60   2 85
  Atropine . . . . 1-60 gr.
  Morphine sulphate . . . . 1-8 gr.
  Dose, 1 to 2.

113. Atropine valerianate, 1-60 gr. . 45   2 10
  Sedative, antispasmodic.
  Dose, 1 to 2.

114. Ballou's 45   2 10
  Comp. Extract Colocynth . . . 1 gr.
  Extract Jalap . . . . 1 gr.
  Mild Mercurous chloride . . . 1 gr.
  Ipecac . . . . 1-8 gr.
  Purgative. Dose, 1 to 3.

Bamboo Brier Comp. See Alterative Comp. (38) 50   2 35
Barclay's. See Anti-Bilious "A"
  (50) 45   2 10
Barker's. See Laxative Special.
  (386) 45   2 10
Bauer's. See Aperient "C" (83) 45   2 10

No.		Per 100.	Per 500.
115.	Belladonna leaves, Ext., 1-8 gr. .	$ .25	$1 10
	Narcotic, sedative. Dose, 1 to 4.		
116.	Belladonna leaves, Ext., 1-4 gr. .	25	1 10
	Dose, 1 to 3.		
117.	Belladonna leaves, Ext., 1-2 gr. .	30	1 35
	Dose, 1 to 2.		
118.	Belladonna leaves, Ext., 1 gr. .	40	1 85
	Dose, 1 to 2.		
119.	Berberine hydrochlorate, 1 gr. .	95	4 60
	Stomachic tonic, antiperiodic.		
	Dose, 1 to 2.		
120.	Berberine sulphate, 2 grs. . . .	1 50	7 35
	Dose, 1 to 2.		
121.	Berberis Aquifolium, Ext., 2 grs.	45	2 10
	Stomachic bitter, tonic.		
	Dose, 1 to 2.		
122.	Bilious "A," Junge's	45	2 10
	Manganese iodide . . . . . . 1-2 gr.		
	Juglandin . . . . 3-10 gr.		
	Extract Hyoscyamus . . . . 3-5 gr.		
	Leptandrin. . . . . 3-10 gr.		
	Sanguinarin . . 1 . . . . . . 1-5 gr.		
	Cholagogue, cathartic.		
	Dose, 1 to 2.		
123.	Bilious "B," Wann's	30	1 35
	Compound Extract Colocynth . 1 gr.		
	Resin Podophyllum . . . . . 1-4 gr.		
	Extract Jalap . . . . 1-4 gr.		
	Extract Hyoscyamus . . . . 1-8 gr.		
	Capsicum . . . . 1-4 gr.		
	Cholagogue, cathartic.		
	Dose, 1 to 2.		
124.	Bismuth and Ignatia	75	3 60
	Bismuth subcarbonate . . . . 4 grs.		
	Extract Ignatia . . . . . . 1-4 gr.		
	Astringent, tonic. Dose, 1 to 2.		
125.	Bismuth and Nux Vomica . . .	80	3 85
	Bismuth subcarbonate . . . . 5 grs.		
	Extract Nux Vomica . . . 1-2 gr.		
	Astringent, tonic. Dose, 1 to 2.		
126.	Bismuth subcarbonate, 3 grs. . .	55	2 60
	Sedative, astringent. Dose, 1 to 3.		
127.	Bismuth subnitrate, 2 grs. . . .	35	1 60
	Dose, 1 to 3.		
128.	Bismuth subnitrate, 3 grs. . . .	40	1 85
	Dose, 1 to 2.		

No.		Per 100.	Per 500.
129.	Black Haw, Extract, 3 grs. . . . $	35	$1 60
	Nervine, tonic. Dose, 1 to 2.		
	Blancard's. See Iron iodide,		
	Blancard (361)	45	2 10
	Blaud's. See Ferrous Carbonate,		
	Blaud's.		
	Blue-mass. See Mercurial.		
	Brown-Sequard's. See Neuralgia, .		
1	"A" (434)	95	4 60
	Brundage's. See Anti-Constipa-		
	tion, "C" (57)	40	1 85
130.	Caffeine, citrated, 1-gr.	40	1 85
	Diuretic, cardiac and nerve stimu-		
	lant. See note 20. 'Dose, 1 to 4.		
131.	Calabar Bean, Ext., 1-12 gr. . . .	35	1 60
	Spinal sedative. Dose, 1 to 2.		
132.	Calabar Bean, Ext., 1-8 gr . . .	35	1 60
	Dose, 1.		
133.	Calcium sulphide, 1-12 gr. . . .	25	1 10
	Alterative and resolvent in skin dis-		
	eases. Dose, 1 to 3.		
134.	Calcium sulphide, 1-10 gr. . . .	30	1 35
	Dose, 1 to 3.		
135.	Calcium sulphide, 1-5 gr	30	1 35
	Dose, 1 to 2.		
136.	Calcium sulphide, 1-4 gr	30	1 35
	Dose, 1 to 2.		
137.	Calcium sulphide, 1-2 gr	30	1 35
	Dose, 1 to 2.		
138.	Calcium sulphide, 1 gr.	35	1 60
	Dose, 1 to 2.		
139.	Calcium sulphide, 1 1-2 grs. . .	35	1 60
	Dose, 1 to 2.		
140.	Calcium sulphide, 2 grs.	40	1 85
	Dose, 1 to 2.		
141.	Calomel, 1-20 gr.	20	85
	Mild Mercurous chloride.		
	Alterative, purgative, antisyphilitic,		
	Dose, 1 to 3.		
142.	Calomel, 1-10 gr.	20	85
	Dose, 1 to 3.		
143.	Calomel, 1-5 gr.	20	85
	Dose, 1 to 3.		

No.   Per 100.   Per 500.

144. Calomel, 1-2 gr. . . . . $1 20 $ 85
    Dose, 1 to 3.

145. Calomel, 1 gr.   20   85
    Dose, 1 to 3.

146. Calomel, 2 grs.   25   1 10
    Dose, 1 to 3.

147. Calomel, 3 grs.   25   1 10
    Dose, 1 to 2.

148. Calomel, 5 grs.   35   1 60
    Dose, 1 to 2.

149. Calomel and Colocynth   45   2 10
    Mild Mercurous chloride . . . 1 gr.
    Comp. Ext. Colocynth . . 2 1-2 grs.
    Cathartic.   Dose, 1 to 2.

150. Calomel and Ipecac   30   1 35
    Mild Mercurous chloride . . . 1 gr.
        Ipecac . . . . 3 1-3 grs.
        Cholagogue, cathartic.
        Dose, 1 to 2.

151. Calomel and Jalap   30   1 35
    Mild Mercurous chloride . . . 1 gr.
    Jalap . . . . 4 grs.
    Cathartic.   Dose, 1 to 2.

152. Calomel, Nux Vomica and Podo-
        phyllin   45   2 10
        Mild Mercurous chloride . . . 1 gr.
        Extract Nux Vomica . . . . 1-2 gr.
        Resin Podophyllum . . . . . 1-2 gr.
        Cholagogue, cathartic.
        Dose, 1 to 2.

153. Calomel and Opium   50   2 35
    Mild Mercurous chloride . . . 2 grs.
    Opium . . . . 1 gr.
    Cathartic.   Dose, 1.

154. Calomel and Rhubarb   40   1 85
    Mild Mercurous chloride . . . 1-2 gr.
    Extract Rhubarb . . . . . . 1-2 gr.
    Comp. Extract Colocynth . . 1-2 gr.
    Extract Hyoscyamus . . . . 1-6 gr.
    Cathartic.   Dose, 1 to 3.

    Calomel and Santonin.   See San-
        tonin and Calomel (604)   55   2 60

155. Camphor, 1 gr.   30   1 35
    Sedative, hypnotic.   Dose, 1 to 3.

No.		Per 100.	Per 500.

156. Camphor Comp. . . . . $.50 . . $2.35
    Camphor . . . . 1 gr.
    Opium . . . . 1 gr.
    Kino . . . . 1 gr.
    Oleoresin Capsicum . . . . . 1-16 gr.
    Antispasmodic, anodyne.
    Dose, 1 to 3.

157. Camphor and Hyoscyamus . . . . 30 . . 1.35
    Camphor . . . . 1 gr.
    Extract Hyoscyamus . . . . . 1 gr.
    Antispasmodic, sedative.
    Dose, 1 to 2.

158. Camphor, Hyoscyamus and Valer-
    ian 35   1.60
    Camphor . . . . 1 gr.
    Extract Hyoscyamus . . . . 1-2 gr.
    Valerian . . . . 1-2 gr.
    Antispasmodic, hypnotic, sedative.
    Dose, 1 to 2.

159. Camphor, Monobromated, 1 gr. .   30   1 35
    Nervous sedative. Dose, 1 to 4.

160. Camphor, Monobromated, 2 grs. . 50   2 35
    Dose, 1 to 2.

161. Camphor, Monobromated, 3 grs. .75   3 60
    Dose, 1 to 2.

162. Camphor and Opium   50   2 35
    Camphor . . . . 2 grs.
    Opium . . . . 1 gr.
    Sedative, astringent. Dose, 1.

163. Camphor, Opium and Hyoscya-
    mus   40   1 85
    Camphor . . . . 1 gr.
    Opium . . . . 1-2 gr.
    Extract Hyoscyamus . . . . . 1 gr.
    Sedative, astringent. Dose, 1 to 2.

164. Camphor, Opium and Lead ace-
    tate   50   2 35
    Camphor . . . . 1 gr.
    Opium . . . . 1 gr.
    Lead acetate . . . . 1 gr.
    Sedative, astringent. Dose, 1.

165. Camphor, Opium and Tannic
    acid   35   1 60
    Camphor . . . . 1 gr.
    Opium . . . . 1-4 gr.
    Tannic acid . . . . 2 grs.
    Sedative, astringent. Dose, 1 to 3.

PILLS.

No.		Per 100.	Per 500.

166. Cannabis Indica, Extract, 1-4 gr. $ 35 $1 60
    Narcotic, hypnotic, sedative.
    Dose, 1 to 2.

167. Cannabis Indica, Extract, 1-2 gr.   55   2 60
    Dose, 1 to 2.

168. Cannabis Indica, Extract, 1 gr. 1   95   4 60
    Dose, 1 2

169. Capsicum, 1 gr.   35   1 60
    Stimulant, carminative, stomachic
    tonic. Dose, 1 to 3.

170. Capsicum Comp., Dr. Andrews'.   40   1 85
    Mild Mercurous chloride . . . 1-4 gr.
    Washed Capsicum . . . . . 1-30 gr.
    Emetine . . . . 1-30 gr.
    Morphine meconate . . . . . 1-30 gr.
    Intestinal analgesic. Dose, 1 to 4.

171. Capsicum Oleoresin, 1-2 gr. . . .   45   2 10
    Stimulant, carminative, stomachic
    tonic. Dose, 1 to 3.

172. Carbolic acid, 1-2 minim . . .   20   85
    Antemetic, antiseptic.
    Dose, 1 to 3.

173. Cascara Cathartic, Dr. Hinckle's. 60   2 85
    Cascarin . . . . 1-4 gr.
    Resin Podophyllum . . . . 1-6 gr.
    Strychnine . . . . 1-60 gr.
    Aloin . . . . 1-4 gr.
    Extract Belladonna . . . . . 1-8 gr.
    Oleoresin Ginger . . . . . 1-8 gr.
    Dose, 1 to 2.

174. Cascara Sagrada, Extract, 1 gr. . 30   1 35
    Tonic laxative. Dose, 1 to 3.

175. Cascara Sagrada, Extract, 2 grs. 40   1 85
    Dose, 1 to 3.

176. Cascara Sagrada, Extract, 3 grs. 50   2 35
    Dose, 1 to 2.

177. Cascara Sagrada Compound . .   60   2 85
    Extract Cascara Sagrada . . . 2 grs.
    Extract Nux Vomica . . . . 1-8 gr.
    Extract Belladonna . . . . 1-16 gr.
    Tonic laxative. Dose, 1 to 3.

No.   Per 100.   Per 500.

178. Catarrh, Hager's . . . . $ 45 $2 10
     Quinidine sulphate. . . . . . 3-8 gr.
     Cinchonidine sulphate . . . . 3-8 gr.
     Althaea . . . . 1-4 gr.
     Gentian . . . . 1-4 gr.
     Red Sandalwood . . . . . . 1-12 gr.
     Hydrochloric acid . . . . . . 1-4 gr.
       Anti-catarrhal. Dose, 1 to 3.

179. Cathartic Comp., U. S. P. . . . 25   1 10
     .08 Gm. Co. Ext.Colocynth 1 1-4 grs.
     .06 Gm. Mild Mercurous chlor. 1 gr.
     .03 Gm. Extract Jalap . . . 1-2 gr.
     .015 Gm. Gamboge . . . . 1-4 gr.
     Dose, 1 to 3.

180. Cathartic Compound, Improved : . 25   1 10
     Compound Extract Colocynth . 1 gr.
     Resin Podophyllum. . . . . 1-4 gr.
     Extract Hyoscyamus . . . . 1-4 gr.
     Extract Jalap . . . . 1-2 gr.
     Extract Gentian . . . . . . 1-2 gr.
     Leptandrin . . . . 1-4 gr.
     Oil Peppermint . . . . q. s.
       Dose, 1 to 3.

181. Cathartic, Vegetable, U. S. P. . . 25   1 10
     .06 Gm. Comp. Ext. Colocynth 1 gr.
     .03 Gm. Ext. Hyoscyamus . 1-2 gr.
     .03 Gm. Ext. Jalap . . . . 1-2 gr.
     .015 Gm. Ext. Leptandra . . 1-4 gr.
     .015 Gm. Resin Podophyllum. 1-4 gr.
     .008 Cc. Oil Peppermint . . . 1-8 m.
     Dose, 1 to 3.

    Cathartic Compound "A," Liver
     (392). 25c   1 10
     Socotrine Aloes . . . . 1 gr.
     Mild Mercurous chloride . . . 1-8 gr.
     Leptandrin . . . . 1-8 gr.
     Tinct. Veratrum Viride . . . 1-4 gr.
     Jalap . . . . 1 gr.
     Gamboge . . . . 1-8 gr.
     Oleoresin Capsicum. . . . . 1-48 gr.
       Dose, 1 to 3.

PILLS.

No.		Per 100.	Per 500.

Cathartic Compound, "B," Mandrake Liver (393) . . . . . $ 25 $1 10
    Socotrine Aloes . . . . . . 1 gr.
    Resin Podophyllum . . . . 1-8 gr.
    Leptandrin . . . . 1-8 gr.
    Tinct. Veratrum Viride . . . 1-4 gr.
    Jalap . . . . 1 gr.
    Gamboge . . . . 1-8 gr.
    Oleoresin Capsicum. . . . . 1-48 gr.
    Dose, 1 to 3.

182. Cathartic Compound, "C," Cholagogue. 30   1 35
    Resin Podophyllum . . . . . 1-2 gr.
    Extract Hyoscyamus . . . . 1-8 gr.
    Oleoresin Capsicum . . . . 1-8 gr.
    Mercury mass . . . . 1-4 gr.
    Extract Nux Vomica . . . . 1-16 gr.
    Dose, 1 to 3.

183. Cathartic Comp., "D," Modified. 30   1 35
    Comp. Extract Colocynth . . . 1 gr.
    Extract Jalap . . . . . . 1-4 gr.
    Mild Mercurous chloride . . . 3-4 gr.
    Gamboge . . . . 1-6 gr.
    Rhubarb . . . . 1-2 gr.
    Ginger . . . . 1-4 gr.
    Dose, 1 to 3.

184. Cathartic Comp., "E," Mild . . . 25   1 10
    Comp. Extract Colocynth . . . 1 gr.
    Extract Jalap . . . . . . 1-2 gr.
    Mild Mercurous chloride . . . 1-2 gr.
    Extract Hyoscyamus . . . . 1-2 gr.
    Gamboge . . . . 1-4 gr.
    Oil Peppermint . . . . . . 1-4 gr.
    Dose, 1 to 3.

185. Cathartic Comp., "F," Vegetable. 30   1 35
    Extract Colocynth . . . . . 1-3 gr.
    Resin Podophyllum . . . . 1-4 gr.
    Resin Scammony . . . . . . 1-3 gr.
    Socotrine Aloes . . . . . 1 1-4 grs.
    Cardamom . . . . 1-8 gr.
    Soap . . . . 1-8 gr.
    Dose, 1 to 3.

186. Cathartic, "G," Purgative . . . 25   1 10
    Socotrine Aloes . . . . . . 1 1-8 gr.
    Gamboge. . . . . 3-16 gr.
    Resin Podophyllum . . . . 1-8 gr.
    Washed Capsicum . . . . . 1-8 gr.
    Croton Oil . . . . 1-50 gr.
    Dose, 1 to 3.

No.   Per 100. Per 500.

187. Cerium oxalate, 1-4 gr. . . . . $1 30 . $1 35
      Stomachic, sedative, antemetic. .
      Dose, 1 to 4.

188. Cerium oxalate, 1 gr. 50   2 35
      Dose, 1 to 2.

189. Chalybeate Comp., Jarvis' Im-
      proved 30   1 35
      Extract Nux Vomica . . . . 1-10 gr.
      Ferrous carb. mass (Blaud's) . 3 grs.
      Dose, 1 to 2.

      Chalybeate, Flint's. See Saline
        Chalybeate-Tonic (597) . . . 40   1 85

      Chapman's Dinner. See Dinner,
        "B" (276) 30   1 35

190. Charcoal, 3 grs. 35   1 60
      Stomachic sedative, antiseptic.
      Dose, 1 to 3.

191. Chinoidine, 2 grs. 25   1 10
      Tonic. Dose, 1 to 2.

192. Chinoidine, 5 grs. 40   1 85
      Dose, 1 to 2.

193. Chinoidine Compound, "A" . . . 45   2 10
      Chinoidine . . . . 2 grs.
      Comp. Extract Colocynth . . 1-3 gr.
      Oleoresin Black Pepper . . 1-6 gr.
      Dried Ferrous sulphate . . 1-2 gr.
      Tonic. Dose, 1 to 2.

194. Chinoidine Comp., "B," Wat-
      son's Anti-Chill 40   1 85
      Chinoidine . . . . 1 gr.
      Iron ferrocyanide . . . . . . 2 grs.
      Oleoresin Black Pepper . . . 1 gr.
      Arsenous acid . . . . . . 1-20 gr.
      Tonic. Dose, 1 to 2.

195. Chinoidine Compound, "C" . . 45   2 10
      Chinoidine . . . . 2 grs.
      Dried Ferrous sulphate . . . 1 gr.
      Oleoresin Black Pepper . . 1-2 gr.
      Tonic. Dose, 1 to 2.

196. Chinoidine Compound, "D" . . 45   2 10
      Chinoidine . . . . 2 grs.
      Dried Ferrous sulphate . . . 1 gr.
      Piperine . . . . 1-2 gr.
      Tonic, stomachic. Dose, 1 to 2.

No.		Per 100.	Per 500.
· · Christopher's. See Liver, "C," (394)		$ 35	$1 60
197. Cimicifugin, 1 gr.		30	1 35
Sedative, diaphoretic. Dose, 1 to 3.			
198. Cinchonidine, 1 gr.		20	85
Tonic. Dose, 1 to 3.			
199. Cinchonidine, 2 grs.		25	1 10
Dose, 1 to 2.			
200. Cinchonidine, 3 grs.		·30	1 35
Dose, 1 to 2.			
201. Cinchonidine salicylate, 1+2 grs.		60	2 85
Dose, 1 to 3.			
202. Cinchonidine sulphate, 1-2 gr. .		20	85
Dose, 1 to 2.			
203. Cinchonidine sulphate, 1 gr. . .		25	1 10
Dose, 1 to 2.			
204. Cinchonidine sulphate, 2 grs. . .		35	1 60
Dose, 1 to 2.			
205. Cinchonidine sulphate, 3 grs. . .		40	1 85
Dose, 1 to 2.			
206. Cinchonidine sulphate, 4 grs. . .		45	2 10
Gelatin-coated only. See note 89. Dose, 1 to 2.			
207. Cinchonidine sulphate, 5 grs. . .		50	2 35
Gelatin-coated only. See note 89. Dose, 1 to 2.			
208. Cinchonidine Comp.		40	1 85
Cinchonidine sulphate . . . . . 1 gr. Arsenous acid . . . . 1-32 gr. Reduced Iron . . . . 1 gr. Ferruginous tonic. Dose, 1 to 2.			
209. Cinchonidine Comp. with Strychnine		40	1 85
Cinchonidine sulphate . . . . . 1 gr. Arsenous acid . . . . 1-20 gr. Reduced Iron . . . . 1 gr. Strychnine . . . . 1-20 gr. Tonic. Dose, 1 to 2.			
210. Cinchonidine and Iron		40	1 85
Cinchonidine sulphate . . . . . 1 gr. Reduced Iron . . . . 1 gr. Tonic. Dose, 1 to 2.			

No.		Per 100.	Per 500.
211.	Cinchonidine, Iron and Strychnine	$ .40	$1 85
	Cinchonidine sulphate . . . . 1 gr.		
	Ferrous carbonate mass . . . 2 grs.		
	Strychnine sulphate . . . . 1-60 gr.		
	Ferruginous tonic. . Dose, 1 to 3.		
212.	Cinchonine, 1 gr.	20	85
	Tonic, antiperiodic. Dose, 1 to 3.		
213.	Cinchonine, 2 grs.	30	1 35
	Dose, 1 to 2.		
214.	Cinchonine, 3 grs.	40	1 85
	Dose, 1 to 2.		
215.	Cinchonine sulphate, 1 gr. . . .	25	1 10
	Dose, 1 to 3.		
216.	Cinchonine sulphate, 1 1-2 grs. .	30	1 35
	Dose, 1 to 3.		
217.	Cinchonine sulphate, 2 grs. . .	35	1 60
	Dose, 1 to 2.		
218.	Cinchonine sulphate, 3 grs. . .	40	1 85
	Dose, 1 to 2.		
219.	Cinchonine sulphate, 5 grs. . . .	50	2 35
	Gelatin-coated only. See note 89.		
	Dose, 1 to 2.		
220.	Cinchonine Compound, "Waxham's Tonic"	40	1 85
	Cinchonine sulphate . . . . 1 gr.		
	Extract Prickly-ash bark . 1-4 gr.		
	Extract Red Cinchona . . . . 1-4 gr.		
	Extract Dogwood . . . . . 1-4 gr.		
	Capsicum . . . . 1-4 g.		
	Tonic, antiperiodic. Dose, 1 to 4.		
221.	Coca, Extract, 2 grs.	50	2 35
	Stimulant. Dose, 1 to 3.		
222.	Coca, Extract, 3 grs.	60	2 85
	Dose, 1 to 2.		
223.	Codeine, 1-16 gr.	45	2 10
	Anodyne. Dose, 1 to 3.		
224.	Codeine, 1-8 gr.	60	2 85
	Dose, 1 to 3.		
225.	Codeine, 1-4 gr.	90	4 35
	Dose, 1 to 2.		
226.	Codeine, 1-2 gr.	1 50	7 35
	Dose, 1 to 2.		
227.	Colchicum seed, Extract, 1-2 gr.	40	1 85
	Antipodagric. Dose, 1 to 3.		

No.   Per 100.   Per 500.

228. Colocynth Compound . . . . . $ 40 $1 85
   Compound Extract Colocynth. 3 grs.
   Dose, 1 to 2.

229. Colocynth Compound, B. P. . .   40   1 85
   Sugar-coated only. See note 90.
   Powd. Colocynth pulp . . . . 1-2 gr.
   Powd. Barbados Aloes . . . . . 1 gr.
   Powd. Scammony resin . . . . 1 gr.
   Potassium sulphate . . . . . . 1-8 gr.
   Oil Cloves . . . . 1-8 gr.
   Cathartic.   Dose, 1 to 3.

230. Colocynth Comp., B. P., 5 grs. .   70   3 35
   Gelatin-coated only.   See note 89.
   Powd. Colocynth pulp . . . 10-11 gr.
   Powd. Scammony resin . . 1 9-11 grs.
   Powd. Barbados Aloes . . 1 9-11 grs.
   Potassium sulphate . . . . . 5-22 gr.
   Oil Cloves . . . . 5-22 gr.
   Cathartic.   Dose, 1 to 2.

   Colocynth Comp. and Blue. See
      Colocynth Comp. and Mercury.

231. Colocynth Comp. and Calomel . .   45   2 10
   Comp. Ext. Colocynth . . . . 2 grs.
   Mild Mercurous chloride . . . 1 gr.
   Cathartic.   Dose, 1 to 2.

232. Colocynth and Hyoscyamus, B.P. .   45   2 10
   Sugar-coated only.   See note 90.
   Compound Pill Colocynth . . 2 grs.
   Extract Henbane . . . . . . 1 gr.
   Cathartic.   Dose, 1 to 2.

**233. Colocynth   Comp.,   Hyoscyamus
   and   Podophyllin,   "A"** . . .   55   2 60
   Comp. Extract Colocynth . . 3 grs.
   Extract Hyoscyamus . . . . . 1 gr.
   Resin Podophyllum . . . . 1-4 gr.
   Cathartic, cholagogue.
   Dose, 1 to 2.

**234. Colocynth   Comp.,   Hyoscyamus
   and   Podophyllin, "B"**   50   2 35
   Comp. Extract Colocynth . . 3 grs.
   Extract Hyoscyamus . . . . 1 gr.
   Resin Podophyllum . . . . 1-8 gr.
   Cathartic, cholagogue.
   Dose, 1 to 2.

No.   Per 100. Per 500.

235. Colocynth Comp. and Mercury,
     "A" . . . . $ 45 $2 10
       Comp. Ext. Colocynth . . 2 1-2 grs.
       Mercury mass . . . . 1-2 gr.
       Cathartic, cholagogue.
       Dose, 1 to 2.

236. Colocynth Comp. and Mercury,
     "B" 50   2 35
       Comp. Extract Colocynth . 2 1-2 grs.
       Mercury mass . . . . . . 2 1-2 grs.
       Cathartic, cholagogue.
       Dose, 1 to 2.

237. Colocynth Compound, Mercury
     and Hyoscyamus 55   2 60
       Comp. Extract Colocynth . . 3 grs.
       Mercury mass . . . . 1 gr.
       Extract Hyoscyamus . . . . . 1 gr.
       Cathartic, cholagogue.
       Dose, 1 to 2.

238. Colocynth Compound, Mercury
     and Ipecac 45.   2 10
       Comp. Extract Colocynth . . 2 grs.
       Mercury mass . . . . 2 grs.
       Ipecac . . . . 1-6 gr.
       Cathartic, cholagogue.
       Dose, 1 to 2.

239. Colocynth Compound, Nux Vom-
     ica and Belladonna 45   2 10
       Comp. Extract Colocynth . . 2 grs.
       Extract Nux Vomica . . . . . 1-4 gr.
       Extract Belladonna . . . . . 1-8 gr.
       Cathartic, tonic. Dose, 1 to 2.

240. Colocynth Compound and Pod-
     ophyllin, "A" 45   2 10
       Comp. Ext. Colocynth . 2 1-2 grs.
       Resin Podophyllum . . . . . 1-4 gr.
       Cathartic, cholagogue.

241. Colocynth Compound and Pod-
     ophyllin, "B" 45   2 10
       Comp. Ext. Colocynth . 2 1-2 grs.
       Resin Podophyllum . . . . 1-2 gr.
       Cathartic, cholagogue.
       Dose, 1 to 2.

242. Conium Extract, 1-2 gr. 30   1 35
       Narcotic, sedative. Dose, 1 to 3.

No.	Per 100.	Per 500.
Cook's. See Anti-Bilious, "B," (51)	$ 25	$1 10
243. Copaiba, 3 grs. . . . . 30 .		1 35
Sugar-coated only. See note 90. Stimulant diuretic. Dose, 1 to 3.		
244. Copaiba, 4. grs. 35		1 60
Sugar-coated only. See note 90. Dose, 1 to 2.		
245. Copaiba Compound "A," Dr. Aiken's 55		2 60

245 ingredients:
```
Copaiba . . . . . . . . 1 1-2 grs.
Oleoresin Cubeb. . , . . . . 1 gr.
Dried Ferrous sulphate . . : 2-3 gr.
Carbolic acid, . . . . . . . 1-3 gr.
Extract Belladonna . . . . . 1-8 gr.
Extract Aconite . . . . . . 1-10 gr.
Oil Peppermint, q. s. . .
Stimulating diuretic, expectorant.
Dose, 1 to 2.
```

| 246. Copaiba Compound "B" . . . . 55 | | 2 60 |

```
Sugar-coated only. See note 90.
Copaiba . . . . . . . . . 1 gr.
Oleoresin Cubeb . . . . . . 1 gr.
Resin Guaiac . . . . . . . 1 gr.
Ferric citrate . . . . . . . 1 gr.
Stimulating diuretic, expectorant.
Dose, 1 to 2.
```

| 247. Copaiba and Cubeb . . . . . 55 | | 2 60 |

```
Sugar-coated only. See note 90.
Copaiba . . . . . . . . . 2 grs.
Oleoresin Cubeb . . . . . . 1 gr.
Stimulating diuretic, expectorant.
Dose, 1 to 2.
```

| 248. Copaiba, Cubeb and Ferric citrate 55 | | 2 60 |

```
Copaiba . . . . . . . . . 1 gr.
Cubeb . . . . . . . . . 1 gr.
Ferric citrate . . . . . . . 1 gr.
Stimulating diuretic, expectorant.
Dose, 1 to 2.
```

249. Cornin, 1-2 gr. 30		1 35
Bitter tonic. Dose, 1 to 3.		
250. Cornin, 1 gr. 40		1 85
Dose, 1 to 3.		
251. Cornin, 2 grs. 45		2 10
Dose, 1 to 2.		

No.		Per 100.	Per 500.
252.	Corrosive Sublimate, 1-100 gr. . .	$ 20	1 $.0 85
	Corrosive Mercuric chloride. . . .		
	Alterative, antisyphilitic.		
	Dose, 1 to 3.		
253.	Corrosive Sublimate, 1-50 gr. . .	20	85
	Dose, 1 to 3.		
254.	Corrosive Sublimate, 1-40 gr. . . .	20 1	85
	Dose, 1 to 3.		
255.	Corrosive Sublimate, 1-32 gr. . .	20	85
	Dose, 1 to 2.		
256.	Corrosive Sublimate, 1-30 gr. . .	20	85
	Dose, 1 to 2.		
257.	Corrosive Sublimate, 1-20 gr. . .	20 1	1 85
	Dose, 1 to 2. . .		
258.	Corrosive Sublimate, 1-16 gr. . .	20	85
	Dose, 1.		
259.	Corrosive Sublimate, 1-12 gr. . .	20	85
	Dose, 1.		
260.	Corrosive Sublimate, 1-8 gr. . .	20	85
	Dose, 1.		
261.	Creosote, Beechwood, 1-16 gr. .	50	2 35
	Antiseptic, stimulant.		
	Dose, 1 to 3.		
262.	Croton Oil, 1-50 minim	30	1 35
	Cathartic. Dose, 1 to 3.		
263.	Croton Oil, 1-3 minim	30	1 35
	Dose, 1 to 2.		
264.	Cubeb, Extract, 2 grs	80	3 85
	Diuretic, expectorant.		
	Dose, 1 to 2.		
265.	Damiana, Extract, 3 grs	55	2 60
	Aphrodisiac. Dose, 1 to 2.		
266.	Damiana Compound	90	4 35
	Extract Damiana, ... 2 grs.		
	Extract Nux Vomica . . . 1-8 gr.		
	Phosphorus. . . . . . 1-100 gr.		
	Aphrodisiac, tonic. Dose, 1 to 2.		
267.	Dandelion, Extract, 3 grs . . .	45	2 10
	Bitter tonic, diuretic. Dose, 1 to 2.		
268.	Dandelion and Leptandrin . . .	30	1 35
	Extract Dandelion . . . . 1 1-3 gr.		
	Leptandrin . . . . . . 2-3 gr.		
	Bitter tonic, diuretic, aperient.		
	Dose, 1 to 2.		

No.		Per 100.	Per 500.

269. Diaphoretic . . . . $ 45 $2 05
- Morphine sulphate . . . . . 1-25 gr.
- Ipecac . . . . 1-4 gr.
- Potassium nitrate . . . . . . 1 gr.
- Camphor
- Dose, 1 to 2.

270. Diarrhea 25   1 10
- Mild Mercurous chloride . . . 1-8 gr.
- Morphine sulphate . . . . 1-16 gr.
- Ipecac . . . . 1-32 gr.
- Capsicum . . . . 1-16 gr.
- Camphor . . . . 1-16 grs.
- Dose, 1 to 3.

271. Digestive, Hager's 50   2 35
- Cinchonidine sulphate . . . 1-5 gr.
- Pepsin (1:3000) . . . . . 1-5 grs.
- Ginger . . . . 3-25 gr.
- Cardamom . . . . 3-25 gr.
- Pimento . . . . 3-25 gr.
- Gentian . . . . 6-25 gr.
- Hydrochloric acid . . . . 1-25 gr.
- Dose, 1 to 3.

272. Digitalin, 1-60 gr. 30   1 35
- Cardiac tonic, diuretic.
- Dose, 1 to 3.

273. Digitalis, Extract, 1-2 grain . . 35   1 60
- Cardiac tonic, diuretic.
- Dose, 1 to 2.

274. Digitalis Compound 30   1 35
- Digitalis . . . . 1 gr.
- Squill . . . . 1 gr.
- Potassium nitrate . . . . . 2 grs.
- Cardiac tonic, diuretic.
- Dose, 1 to 2.

275. Dinner "A," Lady Webster's . . 25   1 10
- Aloes . . . . 4-5 grs.
- Mastic . . . . 3-5 gr.
- Red Rose . . . . 3-5 gr.
- Stimulating laxative. Dose 1 to 2.

276. Dinner "B," Chapman's 30   1 35
- Aloes . . . . 1-4 grs.
- Mastic . . . . 1 gr.
- Rhubarb . . . . 2 grs.
- Stimulating laxative. Dose, 1 to 2.

No.		Per 100.	Per 500.

277. Dipsomania, Mann's . . . . $ 60 $2 85
     Quinine sulphate . . . . 2 grs.
     Strychnine . . . . 1-40 gr.
     Oleoresin Capsicum . . . . . 1-5 gr.
     Zinc oxide . . . . 2 grs.
     Arsenous acid . . . . . . 1-100 gr.
     Dose, 1 to 3.

278. Diuretic 35   1 60
     Digitalis . . . . 1 gr.
     Squill . . . . 1-2 gr.
     Scammony 1 . . . . 1-2 gr.
     Oil Juniper 1 . . . . 1 gr.
     Dose, 1 to 2 . . . .

Dover's Powder. See Ipecac and
   Opium.

Drysdale's. See Aperient "A"
   (81) 35   1 60

279. Dupuytrens Fr. Codex 30   1 35
     Extract Guaiac . . . . 3-5 gr.
     Extract Opium . . . . 3-10 gr.
     Corrosive Mercuric chloride . 1-10 gr.
     Antisyphilitic. Dose, 1 to 2.

280. Dysentery 30   1 35
     Mercury mass . . . . 1 gr.
     Gelsemium . . . . 1-6 gr.
     Ipecac . . . . 1 gr.
     Dose, 1 to 2.

281. Dysmenorrhea, Alvord's 75   3 60
     Morphine sulphate . . . . . 1-10 gr.
     Cimicifugin . . . . 2-3 gr.
     Quinine sulphate . . . . . . 2-3 gr.
     Dose, 1 to 2.

282. Elaterium, Clutterbuck's, 1-10 gr. 50   2 35
     Diuretic, hydragogue cathartic.
     Dose, 1 to 3.

283. Elaterium, Clutterbuck's, 1-8 gr. 55   2 60
     Dose, 1 to 3.

284. Elaterium, Clutterbuck's, 1-4 gr. 1 00   4 85
     Dose, 1.

285. Emmenagogue "A" 70   3 35
     Ergotin, Bonjean's . . . . . 1 gr.
     Extract Black Hellebore . . . 1 gr.
     Aloes . . . . 1 gr.
     Ferrous sulphate . . . . . . 1 gr.
     Oil Savine . . . .
     Dose, 1 to 2.

No.		Per 100.	Per 500.

286. Emmenagogue "B," Hooper's . $ 20 $ 85
Aloes . . . . 1 gr.
Ferrous sulphate . . . . . 1-2 gr.
Extract Black Hellebore . . 1-4 gr.
Myrrh . . . . 1-4 gr.
Canella . . . . 1-8 gr.
Ginger . . . . 1-8 gr.
00 1 Soap . . . . 1-8 gr.
Dose, 1 to 2.

287. Emmenagogue "C," No. 4 . . . 75 3 60
Ergotin . . . . 1 gr.
Extract Cotton-root . . . . . 1 gr.
Purified Aloes . . . . 1 gr.
Dried Ferrous sulphate . . . 1 gr.
Oil Savine. . . . . 1-2 gr.
Dose, 1 to 2.

288. Emmenagogue "D," Mutter's . . 25 1 10
Dried Ferrous sulphate . . 1 1-2 grs.
Turpentine . . . . . . 1 1-2 grs.
25 Aloes . . . . 1 gr.
Dose, 1 to 2.

289. Ergotin, 1-2 gr. 35 1 60
Hemostatic, vasomotor stimulant.
Dose, 1 to 3.

290. Ergotin, 1 gr. 50 2 35
Dose, 1 to 3.

291. Ergotin, 2 grs. 75 3 60
Dose, 1 to 2.

292. Ergotin, 3 grs. 1 00 4 85
Dose, 1 to 2.

293. Ergotin, 5 grs. 1 40 6 85
Gelatin-coated only. See note 89.
Dose, 1.

294. Ergotin Comp., Goodson's . . . 1 50 7 35
Ergotin . . . . 3 grs.
Extract Cannabis Indica . . 1-6 gr.
Strychnine sulphate . . . 1-60 gr.
Vasomotor stimulant, tonic, hemostatic. Dose, 1 to 3.

295. Eucalyptus, Extract, 5 grs. . . . 40 1 85
Antiseptic, diaphoretic.
Dose, 1 to 2.

296. Eucalyptus Compound 50 2 35
Extract Eucalyptus . . . . . 1 gr.
Blood-root . . . . 1-8 gr.
Apocynin . . . . 1-2 gr.
Antiperiodic, diaphoretic.
Dose, 1 to 3.

No.		Per 100.	Per 500.
297.	Euonymin, green, 1 gr.	$ 60	$2 85
	Laxative, alterative. Dose, 1 to 3.		
298.	Euonymin, green, 2 grs.	1 00	4 85
	Dose, 1 to 3.		
299.	Euonymin, green, 3 grs.	1 50	7 35
	Dose, 1 to 2.		
	Ferric citrate. See Iron citrate.		
	Ferrous bromide. See Iron bromide.		
300.	Ferrous carbonate, Blaud's, 1 gr.	20	85
	Hematic tonic. Dose, 1 to 3.		
301.	Ferrous carbonate, Blaud's, 2 grs.	20	85
	Dose, 1 to 3.		
302.	Ferrous carbonate, Blaud's, 3 grs.	25	1 10
	Dose, 1 to 3.		
303.	Ferrous carbonate, Blaud's, 4 grs.	25	1 10
	Dose, 1 to 3.		
304.	Ferrous carbonate, Blaud's, 5 grs.	30	1 35
	Dose, 1 to 3.		
305.	Ferrous carbonate, Blaud's, 6 grs.	30	1 35
	Dose, 1 to 3.		
306.	Ferrous carb. Comp., "A 1," Blaud's	40	1 85
	Ferrous carb. mass (Blaud's) . 5 grs.		
	Socotrine Aloes . . . . 1-8 gr.		
	Extract Nux Vomica . . . . 1-16 gr.		
	Arsenous acid . . . . 1-100 gr.		
	Tonic, hematic. Dose, 1 to 2.		
307.	Ferrous carb. Comp., "A 2," Blaud's	35	1 60
	Ferrous carb. mass (Blaud's) . 3 grs.		
	Socotrine Aloes . . . . 1-8 gr.		
	Extract Nux Vomica . . . 1-16 gr.		
	Arsenous acid . . . . 1-100 gr.		
	Tonic, hematic. Dose, 1 to 3.		
308.	Ferrous carbonate, Improved, "A," Blaud's	25	1 10
	Ferrous carb. mass (Blaud's) . 3 grs.		
	Arsenous acid . . . . . . 1-40 gr.		
	Tonic, hematic. Dose, 1 to 2.		
309.	Ferrous carbonate, Improved, "B," Blaud's	30	1 35
	Ferrous carb. mass (Blaud's) . 5 grs.		
	Arsenous acid . . . . . . 1-40 gr.		
	Tonic, hematic. Dose, 1 to 2.		

No.		Per 100.	Per 500.
310.	Ferrous carbonate, Modified "A," Blaud's	$ 25	$1 10
	Ferrous carb. mass (Blaud's) . 3 grs.		
	Arsenous acid . . . . 1-50 gr.		
	Tonic, hematic. Dose, 1 to 3.		
311.	Ferrous carbonate, Modified "B," Blaud's	30	1 35
	Ferrous carb. mass (Blaud's) . 5 grs.		
	Arsenous acid . . . . 1-50 gr.		
	Tonic, hematic. Dose, 1 to 2.		
312.	Ferrous carbonate, Modified "C," Blaud's	25	1 10
	Ferrous carb. mass (Blaud's) . 3 grs.		
	Arsenous acid . . . . 1-60 gr.		
	Tonic, hematic. Dose, 1 to 3.		
313.	Ferrous carbonate, Modified "D," Blaud's	30	1 35
	Ferrous carb. mass (Blaud's) . 5 grs.		
	Arsenous acid . . . . 1-60 gr.		
	Tonic, hematic. Dose, 1 to 3.		
314.	Ferrous carbonate, Vallet's, 2 grs.	25	1 10
	Hematic tonic. Dose, 1 to 3.		
315.	Ferrous carbonate, Vallet's, 3 grs.	25	1 10
	Dose, 1 to 3.		
316.	Ferrous carbonate, Vallet's, 5 grs.	30	1 35
	Dose, 1 to 2.		
	Ferrous iodide. . See Iron iodide (360)	30	1 35
	Flint's. See Saline Chalybeate (597)	40	1 85
	Francis'. See Triplex Optim. (646)	40	1 85
317.	Galbanum Compound	35	1 60
	Galbanum . . . . 1 1-2 grs.		
	Myrrh. . . . . 1 1-2 grs.		
	Asafetida . . . . 1-2 gr.		
	Antispasmodic, expectorant.		
	Dose, 1 to 2.		
318.	Gamboge Compound, B. P. . . .	30	1 35
	Sugar-coated only. See note 90.		
	Powdered Gamboge . . . . . 3-5 gr.		
	Comp. Cinnamon Powder . . 3-5 gr.		
	Powdered Barbados Aloes . . 3-5 gr.		
	Powdered Hard Soap . . . 1 1-5 grs.		
	Purgative. Dose, 1 to 2.		

No.		Per 100.	Per 500.
319.	Gelsemin, resinoid, 1-16 gr.	$ 30	$1 35
	Antispasmodic, diaphoretic, motor-depressant. Dose, 1 to 3.		
320.	Gelsemin, resinoid, 1-8 gr.	30	1 35
	Dose, 1 to 3.		
321.	Gelsemin, resinoid, 1-4 gr.	45	2 10
	Dose, 1 to 2.		
322.	Gentian, Extract, 2 grs.	30	1 35
	Stomachic tonic, alterative. Dose, 1 to 3.		
323.	Gentian Compound	30	1 35
	Gentian . . . . 2-3 gr.		
	Aloes . . . . 2-3 gr.		
	Rhubarb . . . . 1-3 grs.		
	Oil Caraway . . . . 1-5 gr.		
	Purgative, tonic. Dose, 1 to 3.		
324.	Geranin, 1 gr.	35	1 60
	Astringent, anti-diarrheic. Dose, 1 to 3.		
	Glonoin. See Nitro-glycerin.		
325.	Gonorrhea "A".	30	1 35
	Sugar-coated only. See note 90.		
	Cubeb . . . . 1 1-5 grs.		
	Copaiba . . . . 3-5 gr.		
	Dried Ferrous sulphate . . 3-5 gr.		
	Venice Turpentine . . . . 1 gr.		
	Dose, 1 to 3.		
326.	Gonorrhea "B".	35	1 60
	Sugar-coated only. See note 90.		
	Cubeb . . . . 2 grs.		
	Copaiba . . . . 1 gr.		
	Dried Ferrous sulphate . . 1-2 gr.		
	Venice Turpentine . . . . 1 1-2 grs.		
	Dose, 1 to 2.		
	Goodson's. See Ergotin Comp. (294)	1 50	7 35
	Goss'. See Anti-Constipation, "A" (55)	40	1 85
	Goss'. See Laxative, "C" (385).	40	1 85
	Gross'. See Neuralgia, "C" (436)	1 00	4 85
	Gross', without Morphine. See Neuralgia, "D" (437).	90	4 35
327.	Guarana, 3 grs.	1 50	7 35
	Nerve stimulant. Dose, 1 to 2.		

PILLS.

No.		Per 100.	Per 500
	Hager's Catarrh. See Catarrh, Hager's (178)	$ 45	$2 10
	Hager's Digestive. See Digestive, Hager's (271)	50	2 35
328.	Headache, "A"	60	2 85

Extract Opium . . . . . 1-20 gr.
Extract Belladonna. . . . . . 1-8 gr.
Hyoscyamus. . . . . 1-8 gr.
Lactucarium . . . . 1-4 gr.
Dose, 1 to 3.

| 329. | Headache, "B" | 50 | 2 35 |

Lactinated Pepsin . . . . 1 gr.
Guarana . . . . 1-2 gr.
Sodium bicarbonate . . . . . 1 gr.
Cypripedin. . . . . 1-2 gr.
Dose, 1 to 3.

| 330. | Headache, "C" | 30 | 1 35 |

Irisin . . . . 1-10 gr.
Euonymin . . . . 1-4 gr.
Sanguinarin. . . . . . . 1-20 gr.
Resin Podophyllum . . . . 1-20 gr.
Dose, 1 to 3.

| 331. | Hellebore, Extract, 1 gr. | 40 | 1 85 |

Hydragogue cathartic, emmenagogue. Dose, 1 to 3.

Henbane. See Hyoscyamus.

| 332. | Hepatic, "A" | 60 | 2 85 |

Resin Podophyllum . . . . 1-4 gr.
Leptandrin . . . . 1-2 gr.
Irisin . . . . 1-4 gr.
Extract Nux Vomica . . . . 1-16 gr.
Capsicum . . . . 1-3 gr.
Dose, 1 to 2.

| 333. | Hepatic, "B" | 40 | 1 85 |

Mercury mass). . . . . . . 1 gr.
Comp. Extract Colocynth . . 2-3 gr.
Extract Hyoscyamus . . . . 2-3 gr.
Dose, 1 to 2.

| 334. | Hepatic Compound | 45 | 2 10 |

Gelatin-coated only. See note 89.
Mild Mercurous chloride. . 1-4 gr.
Resin Podophyllum . . . . 1-4 gr.
Comp. Extract Colocynth . 1 1-2 grs.
Extract Belladonna leaves . 1-4 gr.
Extract Nux Vomica . . . 1-2 gr.
Ipecac . . . . 1-8 gr.
Oil Anise . . . . 1-8 gr.
Dose, 1 to 2.

No.		Per 100.	Per 500.
	Hinckle's. See Cascara Cathartic (173)	$ 60	$2 85
	Hooper's. See Emmenagogue, "B" (286)	20	85
	Hubbard's. See Quinine and Aloes Comp. "A" and "B."		
	Hufland's. See Laxative "A" (383)	35	1 60
335.	Hydrastin and Podophyllin . . . Hydrastin, resinoid . . . . . 1-4 gr. Resin Podophyllum . . . . . 1-20 gr. Cholagogue, tonic. Dose, 1 to 3.	50	2 35
336.	Hyoscyamus, Extract, 1-8 gr. . . Sedative, hypnotic. Dose, 1 to 3.	20	85
337.	Hyoscyamus, Extract, 1-4 gr. . . Dose, 1 to 3.	25	1 10
338.	Hyoscyamus, Extract, 1-2 gr. . . Dose, 1 to 3.	30	1 35
339.	Hyoscyamus, Extract, 1 gr. . . . Dose, 1 to 2.	35	1 60
340.	Ignatia, Extract, 1-4 gr. . . . . Nerve tonic, stomachic. Dose, 1 to 3.	30	1 35
341.	Ignatia, Extract, 1-2 gr. . . . . Dose, 1 to 3.	35	1 60
342.	Ignatia, Extract, 1 gr. . . . . . Dose, 1 to 2.	60	2 85
343.	Iodoform, 1 gr. . . . . Antiseptic, alterative. Dose, 1 to 3.	60	2 85
344.	Iodoform and Iron . . . . . . Iodoform . . . . 1 gr. Reduced Iron . . . . 1 1-4 grs. Tonic, alterative, antiseptic. Dose, 1 to 2.	70	3 35
345.	Iodoform, Iron and Quinine . . Iodoform . . . . 1 gr. Vallet's mass . . . . 2 grs. Quinine sulphate . . . . . 1-2 gr. Tonic, alterative, antiseptic. Dose, 1 to 2.	90	4 35
346.	Ipecac, Extract, 1-4 gr. . . . . Cholagogue, expectorant. Dose, 1 to 2.	30	1 35

No.	Per 100.	Per 500.

347. Ipecac and Opium, "A" . . . . $ 30 $1 35
    Ipecac . . . . 1-4 gr.
    Opium . . . . 1-4 gr.
    Representing 2 1-2 grs. Dover's
    Powder. Diaphoretic, soporific,
    anodyne. Dose, 1 to 2.

348. Ipecac and Opium, "B" . . . 40   1 85
    Ipecac . . . . 7-20 gr.
    Opium . . . . 7-20 gr.
    Representing 3 1-2 grs. Dover's
    Powder. Dose, 1 to 3.

349. Ipecac and Opium, "C" 50   2.35
    Ipecac . . . . 1-2 gr.
    Opium . . . . 1-2 gr.
    Representing 5 grs. Dover's Pow-
    der. Dose, 1 to 2.

350. Ipecac and Opium, "D" 90   4 35
    Ipecac . . . . 1 gr.
    Opium . . . . 1 gr.
    Representing 10 grs. Dover's Pow-
    der. Dose, 1.

351. Irisin, 1-2 gr. 40   1.85
    Purgative, diuretic. Dose, 1 to 3.

352. Irisin, 1 gr. 45   2 10
    Dose, 1 to 2.

353. Irisin Compound 30   1 35
    Irisin . . . . 1-4 gr.
    Resin Podophyllum . . . . 1-10 gr.
    Strychnine . . . . 1-40 gr.
    Purgative, tonic. Dose, 1 to 3.

354. Iron arsenate, 1-10 gr. 45   2 10
    Tonic, alterative. Dose, 1 to 2.

355. Iron bromide, 1-2 gr. 45   2 10
    Hematic tonic. Dose, 1 to 3.

356. Iron bromide, 1 gr. 45   2 10
    Dose, 1 to 3.
    Iron by hydrogen. See Iron, Re-
    duced.
    Iron carbonate, Blaud's. See
    Ferrous carbonate, Blaud's.
    Iron carbonate, Vallet's. See
    Ferrous carbonate, Vallet's.

357. Iron citrate, 2 grs. 30   1 35
    Hematic tonic. Dose, 1 to 2.

No.		Per 100.	Per 500.

358. Iron Comp., U. S. P. '80 . . . . . $ 25 |$1 10
   Myrrh . . . . 1 1-2 grs.
   Sodium carbonate . . . . . 3-4 gr.
   Ferrous sulphate . . . . . . 3-4 gr.
   Hematic tonic. Dose, 1 to 2.

359. Iron ferrocyanide, 3 grs. . . . . 30 . 1 L 35
   Sedative, hematic tonic.

360. Iron iodide, U. S. P. . . . . 130 . 1 35
   .065 Gm. Ferrous iodide . . . . 1 gr.
   .012 Gm. Reduced Iron . . . 1-5 gr.
   Tonic, alterative. Dose, 1 to 2.

361. Iron iodide, Blancard's; tolu-
   coated 45   2 10
   Ferrous iodide . . . . 1 gr.
   Dose, 1 to 2.

362. Iron iodide and Quinine sulphate. 50   2 35
   Ferrous iodide . . . . 1 1-2 grs.
   Quinine sulphate . . . . . . 1-4 gr.
   Tonic, alterative. Dose, 1 to 2.

363. Iron and Manganese 45   2 10
   Ferrous carb. mass, Vallet's : 2 grs.
   Manganese carbonate . . . . . 1 gr.
   Hematic tonic. Dose, 1 to 2.

   Iron and Myrrh. See Iron Compound (358) 25   1 10

364. Iron phosphate, 2 grs. 40   1 85
   Hematic tonic, nervine. Dose, 1 to 2.

365. Iron and Potassium tartrate, 2 grs. 40   1 85
   Hematic tonic. Dose, 1 to 2.

366. Iron, Quassia and Nux Vomica . 45   2 10
   Reduced Iron . . . . 1 1-2 grs.
   Extract Quassia . . . . . . 1 gr.
   Extract Nux Vomica . . . : 1-4 gr.
   Soap . . . . 1-2 gr.
   Bitter tonic. Dose, 1 to 2.

367. Iron and Quinine citrate, 1 gr . 20   85
   Bitter tonic. Dose, 1 to 3.

368. Iron and Quinine citrate, 2 grs. . 30   1 35
   Dose, 1 to 3.

369. Iron and Quinine citrate, 3 grs. . 40   1 85
   Dose, 1 to 2.

370. Iron and Quinine citrate, 5 grs. . 60   2 85
   Dose, 1 to 2.

PILLS.

No.		Per 100.	Per 500.
371.	Iron, Quinine and Strychnine cit-rates	$ 55	$2 60
	Iron and Quinine citrate . . . 2 grs.		
	Strychnine citrate . . . . . 1-60 gr.		
	Bitter tonic. Dose, 1 to 2.		
372.	Iron, Reduced, 1 gr.	20	85
	Chalybeate. Dose, 1 to 2.		
373.	Iron, Reduced, 2 grs.	35	1 60
	Dose, 1 to 2.		
374.	Iron and Strychnine	45	2 10
	Reduced Iron. . . . . . . 2 grs.		
	Strychnine . . . . 1-60 gr.		
	Bitter tonic, nervine. Dose, 1 to 2		
375.	Iron and Strychnine citrate . . .	45	2 10
	Iron and Ammonium citrate . . 2 grs.		
	Strychnine . . . . 1-50 gr.		
	Bitter tonic, nervine. Dose, 1 to 3.		
376.	Iron sulphate, Exsiccated, 1 gr. .	25	1 10
	Chalybeate. Dose, 1 to 3.		
377.	Iron sulphate, Exsiccated, 2 grs.	25	1 10
	Dose, 1 to 3.		
378.	Iron sulphate, Exsiccated, 4 grs.	25	1 10
	Dose, 1 to 2.		
379.	Iron valerianate, 1 gr.	50	2 35
	Chalybeate, nervine. Dose, 1 to 2,		
380.	Jaborandi, Extract, 3 grs. . . .	50	2 35
	Diaphoretic, sialagogue.		
	Dose, 1 to 2.		
381.	Jalap, Resin, 1 gr.	60	2 85
	Hydragogue cathartic.		
	Dose, 1 to 3.		
382.	Jamaica Dogwood, Extract, 2 grs. .	80	3 85
	Nervous sedative, anodyne.		
	Dose, 1 to 2.		
	Jarvis'. See Chalybeate Comp.		
	(189)	30	1 35
	Junge's. See Bilious "A" (122)	45	2 10
	Kermes Mineral. See Antimony,		
	Sulphurated.		
	Lady Webster's. See Dinner		
	"A" (275)	25	1 10

No.		Per 100.	Per 500.

383. Laxative, "A," Hufland's . . . $ 35 $1 60
Extract Ox-gall . . . . . . 3-4 gr.
Soap . . . . 3-4 gr.
Rhubarb. . . . . 3-4 gr.
Extract Dandelion . . . . . 3-4 gr.
Dose, 1 to 3.

384. Laxative, "B," Warner's 35 1 60
Aloes . . . . 1 gr.
Sulphur . . . . 1-5 gr.
Resin Podophyllum . . . . . 1-5 gr.
Guaiac. . . . . 2-5 gr.
Syrup Buckthorn . . . . . 1-5 gr.
Dose, 1 to 3.

385. Laxative, "C," Dr. Goss' . . . 40 1 85
Euonymin . . . . 1-4 gr.
Resin Podophyllum . . . . . 1-4 gr.
Extract Hyoscyamus . . . . 1-4 gr.
Apocynum. . . . . 1-2 gr.
Compound Ext. Colocynth . . 1-4 gr.
Extract Butternut . . . . . 1-2 gr.
Extract Cascara Sagrada . . . 1-2 gr.
Dose, 1 to 2.

386. Laxative. Special, Fordyce Bar-
ker's 45 2 10
Comp. Extract Colocynth 2-3 grs.
Extract Hyoscyamus . . 1-4 grs.
Purified Aloes. . . . . . . 5-6 gr.
Extract Nux Vomica . . . . 5-12 gr.
Resin Podophyllum . . . . 1-12 gr.
Ipecac . . . . 1-12 gr.
Dose, 1 to 2.

387. Laxatonic 35 1 60
Aloin . . . . 1-4 gr.
Extract Belladonna . . . . . 1-8 gr.
Strychnine . . . . 1-60 gr.
Ipecac . . . . 1-16 gr.
Tonic, laxative. Dose, 1 to 3.

388. Leptandrin, 1-4 gr. 20 85
Tonic, cholagogue. Dose, 1 to 2.

389. Leptandrin, 1-2 gr. 25 1 10
Dose, 1 to 2.

390. Leptandrin Comp.. . . . . . 55 2 60
Leptandrin . . . . 1 gr.
Irisin . . . . 1-4 gr.
Resin Podophyllum . . . . 1-8 gr.
Cholagogue, laxative. Dose, 1 to 2.

PILLS.

No.		Per 100.	Per 500.
	Leptandrin and Podophyllin. (See Podophyllin and Leptandrin (521)	45	2 10
391.	Leucorrhea .... $ 60		$2 85
	Hamamelin .... 2 grs.		
	Hydrastin .... 1-2 gr.		
	Senecin .... 1-2 gr.		
	Dose, 1 to 3.		
392.	Liver, "A" 25		1 10
	Purified Aloes .... 1 gr.		
	Jalap .... 1 gr.		
	Gamboge .... 1-8 gr.		
	Leptandrin .... 1-8 gr.		
	Mild Mercurous chloride .... 1-8 gr.		
	Oleoresin Capsicum .... 1-48 gr.		
	Tincture Veratrum Viride. .... 1-4 gr.		
	Dose, 1 to 3.		
393.	Liver, "B," Mandrake 25		1 10
	Purified Aloes .... 1 gr.		
	Powdered Jalap .... 1 gr.		
	Powdered Gamboge .... 1-8 gr.		
	Leptandrin .... 1-8 gr.		
	Resin Podophyllum .... 1-8 gr.		
	Oleoresin Capsicum .... 1-48 gr.		
	Tincture Veratrum Viride. .... 1-4 gr.		
	Dose, 1 to 2.		
394.	Liver, "C," Christopher's 35		1 60
	Mild Mercurous chloride .... 2 grs.		
	Rhubarb .... 1 gr.		
	Powdered Ipecac .... 1-2 gr.		
	Dose, 1 to 2.		
395.	Liver, "D," Dr. Taylor's 35		1 60
	Resin Podophyllum .... 1 gr.		
	Powdered Ipecac .... 1 gr.		
	Camphor .... 4-5 gr.		
	Dose, 1 to 2.		
396.	Liver, "E," Waxham's 30		1 35
	Extract Leptandra. .... 1 gr.		
	Capsicum .... 1-4 gr.		
	Extract Hyoscyamus .... 1-8 gr.		
	Resin Podophyllum .... 1-4 gr.		
	Extract Jalap .... 1-4 gr.		
	Extract Gentian .... 1-8 gr.		
	Dose, 1 to 2.		

No.		Per 100.	Per 500.

397. Liver, "F," Little . . . . $ 25 '$1 10
    Aloin . . . . 1-10 gr.
    Resin Jalap . . . . 1-10 gr.
    Resin Podophyllum . . . . . 1-5 gr.
    Extract Hyoscyamus . . . . 1-20 gr.
    Extract Nux Vomica . . . . 1-20 gr.
    Oleoresin Capsicum . . . . . 1-20 gr.
    Dose, 1 to 3.

398. Lupulin, 3 grs. 35   1 60
    Tonic. Dose, 1 to 2.
  Maddin's. See Anti-Malarial,
    "C," "D," "E," "F."

399. Manganese dioxide, 1 gr. . . .   25   1 10
    Emmenagogue, hematic tonic.
    Dose, 1 to 3.

400. Manganese dioxide, 2 grs.   40   1 85
    Dose, 1 to 2.
  Mann's. See Dipsomania (277) . . 60   2 85
  McCaw's Anti-Malarial. 'See
    Anti-Malarial "B" (64) . . .   65   3 10

401. Mercurial, 1-2 gr.   20   85
    Mercury mass; Blue mass.
    Alterative, purgative, antisyphilitic.
    Dose, 1 to 3:

402. Mercurial, 1 gr.   20   85
    Dose, 1 to 3.

403. Mercurial, 2 1-2 grs.   20   85
    Dose, 1 to 2.

404. Mercurial, 3 grs.   20   85
    Dose, 1 to 2.

405. Mercurial, 5 grs.   25   1 10
    Dose, 1 to 2.

406. Mercurial Compound   45   2 10
    Mercury mass . . . . 1 gr.
    Opium . . . . 1-2 gr.
    Ipecac . . . . 1-4 gr.
    Alterative, purgative, antisyphilitic.
    Dose, 1 to 2.
  Mercuric chloride. See Corrosive
    Sublimate.

407. Mercuric iodide, 1-40 gr. . . .   20   85
    Red or Deuto-iodide of Mercury.
    Alterative, tonic. Dose, 1 to 3.

408. Mercuric iodide, 1-25 gr. . . .   20   85
    Dose, 1 to 3.

No,		Per 100.	Per 500.
409.	Mercuric iodide, 1-16 gr.	$ 20	$ 85
	Dose, 1.		
410.	Mercuric iodide, 1-10 gr.	20	85
	Dose, 1.		
411.	Mercuric iodide, 1-8 gr.	20	85
	Dose, 1.		
412.	Mercurous iodide, yellow, 1-8 gr.	20	85
	Mercury protiodide.		
	Alterative, antisyphilitic.		
	Dose, 1 to 3.		
413.	Mercurous iodide, yellow, 1-6 gr.	20	85
	Dose, 1 to 3.		
414.	Mercurous iodide, yellow, 1-5 gr.	20	85
	Dose, 1 to 3.		
415.	Mercurous iodide, yellow, 1-4 gr.	20	85
	Dose, 1 to 3.		
416.	Mercurous iodide, yellow, 1-3 gr.	25	1 10
	Dose, 1 to 2.		
417.	Mercurous iodide, yellow, 1-2 gr.	25	1 10
	Dose, 1 to 2.		

Mercurous iodide and Opium.
See Syphilitic, A and B.

418.	Mercurous tannate, 1 gr.	40	1 85
	Gelatin-coated only. See note 89.		
	Antisyphilitic. Dose, 1.		

Mercury bichloride. See Corrosive Sublimate.

Mercury protiodide. See Mercurous iodide, yellow.

419.	Migraine	50	2 35
	Acetanilid . . . . 2 grs.		
	Monobromated Camphor . . . 1-2 gr.		
	Dose, 1 to 3.		

Monobromated Camphor. See Camphor, Monobromated.

420.	Morphine acetate, 1-8 gr.	35	1 60
	Anodyne, soporific. Dose, 1 to 2.		
421.	Morphine acetate, 1-4 gr.	50	2 35
	Dose, 1 to 2.		

Morphine and Atropine, "A,"
(III). 50   2 35
Atropine. . . . . 1-300 gr.
Morphine sulphate . . . . . . 1-24 gr.
Narcotic, anodyne.   Dose, 1 to 3.

No.		Per 100.	Per 500.
	Morphine and Atropine "B," (112).	$ 60	$2 85
	Atropine . . . . 1-60 gr.		
	Morphine sulphate . . . . . . 1-8 gr.		
	Narcotic, anodyne. Dose, 1 to 2.		
422.	Morphine Compound	75	3 60
	Mild Mercurous chloride . . . . 1-4 gr.		
	Morphine sulphate . . . . . . 1-4 gr.		
	Tartar Emetic . . . . 1-4 gr		
	Sedative, anodyne. Dose, 1 to 2.		
423.	Morphine hydrochlorate, 1-6 gr.	35	1 60
	Anodyne, soporific. Dose, 1 to 2.		
424.	Morphine hydrochlorate, 1-4 gr.	50	2 35
	Dose, 1 to 2.		
425.	Morphine sulphate, 1-32 gr.	20	85
	Anodyne, soporific. Dose, 1 to 3.		
426.	Morphine sulphate, 1-20 gr.	20	85
	Dose, 1 to 3.		
427.	Morphine sulphate, 1-16 gr.	20	85
	Dose, 1 to 3.		
428.	Morphine sulphate, 1-10 gr.	20	85
	Dose, 1 to 3.		
429.	Morphine sulphate, 1-8 gr.	25	1 10
	Dose, 1 to 2.		
430.	Morphine sulphate, 1-6 gr.	30	1 35
	Dose, 1.		
431.	Morphine sulphate, 1-4 gr.	35	1 60
	Dose, 1.		
432.	Morphine sulphate, 1-2 gr.	60	2 85
	Dose, 1.		
433.	Morphine valerianate, 1-8 gr.	50	2 35
	Anodyne, sedative. Dose, 1 to 2.		
	Mutter's. See Emmenagogue.		
	"D," (288) . . . . 1a.	25	1 10
434.	Neuralgia "A," Idiopathic, Brown-Sequard's	95	4 60
	Extract Hyoscyamus . . . 2-3 gr.		
	Extract Conium . . . . . . 2-3 gr.		
	Extract Ignatia . . . . . 1-2 gr.		
	Extract Opium . . . . . . 1-2 gr.		
	Extract Aconite . . . . . 1-3 gr.		
	Extract Cannabis Indica . . 1-4 gr.		
	Extract Stramonium . . . . 1-6 gr.		
	Extract Belladonna . . . . 1-6 gr.		
	Dose, 1 to 2.		

No.		Per 100.	Per 500.
435.	Neuralgia "B"	$ 60	$2 85

    Extract Belladonna . . . . . 1-8 gr.
    Morphine sulphate . . . . . . 1-6 gr.
    Strychnine . . . . 1-60 gr.
    Dose, 1 to 2.

436. Neuralgia "C," Dr. Gross', with
    Morphine 1 00   4 85
    Quinine sulphate . . . . 2 grs.
    Arsenous acid . . . . 1-20 gr.
    Extract Aconite. . . . . . . 1-2 gr.
    Morphine sulphate . . . . . . 1-20 gr.
    Strychnine . . . . 1-30 gr.
    Dose, 1 to 2.

437. Neuralgia "D," Dr. Gross', with-
    out Morphine 90   4 35
    Quinine sulphate . . . . 2 grs.
    Arsenous acid . . . . 1-20 gr.
    Extract Aconite. . . . . . . 1-2 gr.
    Strychnine . . . . 1-30 gr.
    Dose, 1 to 2.

438. Night Sweat 55   2 60
    Zinc oxide . . . . 1-2 gr.
    Salicin. . . . . 1 gr.
    Extract Belladonna . . . . . 1-20 gr.
    Hydrastin, resinoid . . . . . . 1 gr.
    Lactinated Pepsin . . . . . 1-2 gr.
    Dose, 1 to 3.

439. Nitro-glycerin, 1-100 gr. 50   2 35
    Cardiac stimulant, spinal depres-
    sant. Dose, 1 to 2.

440. Nitro-glycerin, 1-50 gr. 50   2 35
    Dose, 1 to 2.

441. Nux Vomica, Extract, 1-50 gr. . . 20   85
    Bitter tonic, nervine. Dose, 1 to 3.

442. Nux Vomica, Extract, 1-8 gr. . . 20   85
    Dose, 1 to 3.

443. Nux Vomica, Extract, 1-4 gr. . . 20   85
    Dose, 1 to 2.

444. Nux Vomica, Extract, 1-2 gr. . . 25   1 10
    Dose, 1 to 2.

Oil, Croton. See Croton Oil.

445. Opium, 1-4 gr. 25   1 10
    Narcotic, sedative, anodyne, astrin-
    gent. Dose, 1 to 3.

446. Opium, 1-2 gr. 25   1 10
    Dose, 1 to 2.

No.		Per 100.	Per 500.
447.	Opium, U. S. P. (.065 Gm.), 1 gr.	$ .40	$1 85

Dose, 1.

Opium. See also Soap Compound,
B. P. (607) 35   1 60

| 448. | Opium, Crystal, 1-4 gr. . . . . . | 30 | 1 35 |

Dose, 1 to 3.

| 449. | Opium, Crystal, 1-2 gr. . . . . . | 45 | 2 10 |

Dose, 1 to 2.

| 450. | Opium, Crystal, 1 gr. . . . . . | 75 | 3 60 |

Dose, 1.

| 451. | Opium, Extract, 1-4 gr. . . . . | 40 | 1 85 |

Dose, 1 to 3.

| 452. | Opium, Extract, 1-2 gr. . . . . . | 55 | 2 60 |

Dose, 1 to 2.

| 453. | Opium, Extract, 1 gr. . . . . . . | 75 | 3 60 |

Dose, 1.

454. Opium and Lead acetate "A" .   40   1 85
  Sugar-coated only. See note 90.
  Opium . . . . 1 gr.
  Lead acetate . . . . 1 gr.
  Astringent, sedative. Dose, 1 to 2.

455. Opium and Lead acetate "B" .   35   1 60
  Sugar-coated only. See note 90.
  Opium . . . . 1-2 gr.
  Lead acetate . . . . 1 1-2 grs.
  Astringent, sedative.
  Dose, 1 to 2.

456. Ox Gall   35   1 60
  Ox-gall. . . . . 2 grs.
  Ginger . . . . 1 gr.
  Laxative, antiseptic. Dose, 1 to 3.

457. Ox Gall Comp., Dr. Andrews' .   55 1   2 60
  Ox-gall, inspissated . . . . . 2 grs.
  Extract Stramonium . . . . . 1-6 gr.
  Purified Aloes . . . . 1-10 gr.
  Hydrastine hydrochlorate. 1-12 gr.
  Tonic, antiseptic, purgative.
  Dose, 1 to 2.

Ox Gall Comp. See also Laxa-
  tive "A" (383) 35   1 60
Palmer's. See Anti-Constipation
  "B" (56). 35   1 60

| 458. | Pepsin, Pure, 1-3000, 1 gr. . . | 40 | 1 85 |

Digestive. Dose, 1 to 2.

No.		Per 100.	Per 500.

459. **Pepsin and Bismuth** . . . . . . $.75 $3.60
    Pure Pepsin . . . . . . 1 1-2 grs.
    Bismuth subcarbonate . . . . 3-4 gr.
    Digestive, stomachic sedative.
    Dose, 1 to 2.

460. **Pepsin, Bismuth and Strychnine** 1.00   4 85
    Gelatin-coated only. See note 89.
    Pure Pepsin . . . . 2 1-2 grs.
    Bismuth subnitrate . . . . . . 2 grs.
    Strychnine . . . . 1-60 gr.
    Digestive, stomachic sedative.
    Dose, 1 to 2.

461. **Pepsin and Iron "A"** 55   2 60
    Pure Pepsin . . . . 1 1-2 grs.
    Reduced Iron . . . . 3-4 gr.
    Tonic, digestive. Dose, 1 to 3.

462. **Pepsin and Iron "B"** 75   3 60
    Pure Pepsin . . . . 1 1-2 grs.
    Ferrous iodide . . . . . . 3-4 gr.
    Reduced Iron . . . . . . 1 1-2 grs.
    Tonic, digestive. Dose, 1 to 2.

463. **Petroleum Compound** 35   1 60
    Petroleum mass . . . . 1 gr.
    Tar . . . . 1 gr.
    Iron phosphate . . . . . . 3-4 gr.
    Extract Nux Vomica . . . . 1-8 gr.
    Tonic, antiseptic. Dose, 1 to 2.

464. **Phenacetine-Bayer, 2 grs.** . . . 1 00   4 85
    Analgesic, antipyretic.
    Dose, 1 to 3.

465. **Phenacetine-Bayer, 4 grs.** . . . 1 80   8 85
    Dose, 1 to 3.

466. **Phenacetine and Quinine Comp.** 2 00   9 85
    Gelatin-coated only. See note 89.
    Phenacetine-Bayer . . . . . . 3 grs.
    Dover's Powder . . . . . . 1-2 gr.
    Quinine sulphate . . . . . . 2 grs.
    Extract Aconite root . . . . 1-12 gr.
    Analgesic, sedative, antiperiodic.
    Dose, 1 to 2.

467. **Phenacetine and Salol** 1 85   9 10
    Phenacetine-Bayer . . . . 2 1-2 grs.
    Salol . . . . 2 1-2 grs.
    Analgesic, antipyretic, intestinal
       antiseptic. Dose, 1 to 3.

468. **Phosphorus, 1-100 gr.** 20   85
    Nervine, nutrient to osseous tissue.
    Dose, 1 to 2.

No.   Per 100.   Per 500.

469. Phosphorus, 1-60 gr. . . . . . . $ 20 $ 85
     Nervine, nutrient to osseous tissue.
     Dose, 1 to 2.

470. Phosphorus, 1-50 gr. 20   85
     Dose, 1 to 2.

471. Phosphorus, 1-33 gr. 20   85
     Dose, 1.

472. Phosphorus, 1-25 gr. 20   85
     Dose, 1.

473. Phosphorus, 1-20 gr. 20   85
     Dose, 1.

474. Phosphorus and Aconite 25   1 10
     Phosphorus . . . . 1-50 gr.
     Extract Aconite . . . . 1-16 gr.
     Anti-neuralgic, nervine.
     Dose, 1 to 2.

475. Phosphorus, Aloes, Iron and
     Strychnine 45   2 10
     Phosphorus . . . . 1-50 gr.
     Extract Aloes . . . . 1 gr.
     Ferrous sulphate . . . . . . 1 1-2 grs.
     Strychnine . . . . 1-30 gr.
     General tonic. Dose, 1 to 2.

476. Phosphorus, Aloes and Nux Vom-
     ica, "A" 30   1 35
     Phosphorus . . . . 1-50 gr.
     Extract Aloes . . . . 1-2 gr.
     Extract Nux Vomica . . . . 1-4 gr.
     General tonic. Dose, 1 to 2.

477. Phosphorus, Aloes and Nux Vom-
     ica, "B" 30   1 35
     Phosphorus . . . . 1-20 gr.
     Extract Aloes . . . . 1-2 gr.
     Extract Nux Vomica . . . . 1-2 gr.
     General tonic. Dose, 1 to 2.

478. Phosphorus and Belladonna, "A". 25   1 10
     Phosphorus . . . . 1-100 gr.
     Extract Belladonna . . . . . 1-8 gr.
     Nervine, tonic. Dose, 1 to 2.

479. Phosphorus and Belladonna, "B". 30   1 35
     Phosphorus . . . . 1-33 gr.
     Extract Belladonna . . . . . 1-4 gr.
     Nervine, tonic. Dose, 1 to 2.

PILLS.

No.   Per 100.   Per 500.

480. Phosphorus and Cannabis Indica,
"A" . . . . $ 40 $1 85
Phosphorus . . . . 1-50 gr.
Extract Cannabis Indica . . . 1-4 gr.
Nervine, sedative. Dose, 1 to 2.

481. Phosphorus and Cannabis Indica,
"B"   40   1 85
Phosphorus . . . . 1-25 gr.
Extract Cannabis Indica . . . 1-4 gr.
Nervine, sedative. Dose, 1.

482. Phosphorus and Cantharides Com-
pound, "A"   45   2 10
Phosphorus . . . . 1-50 gr.
Cantharides . . . . 1 gr.
Extract Nux Vomica . . . . . 1 gr.
Sexual invigorator, nerve tonic.
Dose, 1 to 2.

483. Phosphorus and Cantharides Com-
pound, "B"   45   2 10
Phosphorus . . . . 1-33 gr.
Cantharides . . . . 1 gr.
Extract Nux Vomica . . . . . 1 gr.
Sexual invigorator, nerve tonic.
Dose, 1.

484. Phosphorus Comp., "A" 25   1 10
Phosphorus . . . . 1-100 gr.
Extract Nux Vomica . . . . 1-4 gr.
Nerve tonic. Dose, 1 to 3.

485. Phosphorus Comp., "B"   25   1 10
Phosphorus . . . . 1-60 gr.
Extract Nux Vomica . . . . 1-4 gr.
Nerve tonic. Dose, 1 to 3.

486. Phosphorus Comp., "C"   25   1 10
Phosphorus . . . . 1-33 gr.
Extract Nux Vomica . . . . 1-2 gr.
Nerve tonic. Dose, 1.

487. Phosphorus, Damiana and Can-
tharides   75   3 60
Phosphorus . . . . 1-100 gr.
Extract Nux Vomica . . . . 1-8 gr.
Extract Damiana . . . . . 1 gr.
Extract Coca . . . . 1 gr.
Cantharides . . . . 1-2 gr.
Aphrodisiac. Dose, 1 to 3.

No.		Per 100.	Per 500.

Phosphorus, Damiana and Cantharides. See Aphrodisiac Imp.
(84) . . . . $ 75 · $3 60

488. Phosphorus, Digitalis and Henbane "A" . . . . 40 · 1 85
    Phosphorus . . . . 1-50 gr.
    Digitalis . . . . 1 gr.
    Extract Hyoscyamus . . . . 1 gr.
    Nervine, tonic. Dose, 1 to 2.

489. Phosphorus, Digitalis and Henbane "B" 50 2 35
    Phosphorus . . . . 1-33 gr.
    Digitalis . . . . 1 gr.
    Extract Hyoscyamus . . . . 2 grs.
    Nervine, tonic. Dose, 1.

490. Phosphorus, Digitalis and Iron
"A" 45 2 10
    Phosphorus . . . . 1-50 gr.
    Digitalis . . . . 1 gr.
    Reduced Iron . . . . 1 gr.
    Tonic, nervine. Dose, 1 to 2.

491. Phosphorus, Digitalis and Iron
"B" 50 2 35
    Phosphorus . . . . 1-33 gr.
    Digitalis . . . . 1 gr.
    Reduced Iron . . . . 3 grs.
    Tonic, nervine. Dose, 1 to 2.

492. Phosphorus and Iron "A" . . . 30 1 35
    Phosphorus . . . . 1-100 gr.
    Vallet's mass . . . . 1 gr.
    Tonic, nervine. Dose, 1 to 3.

493. Phosphorus and Iron "B" . . . 40 1 85
    Phosphorus . . . . 1-50 gr.
    Reduced Iron . . . . 2 grs.
    Tonic, nervine. Dose, 1 to 2.

494. Phosphorus and Iron "C" . . . 40 1 85
    Phosphorus . . . . 1-50 gr.
    Reduced Iron . . . . 3 grs.
    Tonic, nervine. Dose, 1 to 2.

Phosphorus, Iron and Quinine.
See Quinine, Iron and Phosphorus.

Phosphorus, Iron, Quinine and
Strychnine. See Quinine, Iron,
Phosphorus and Strychnine.

PILLS.

No.   Per 100.   Per 500.

495. Phosphorus, Morphine and Zinc
     valerianate, "A" $1 00 $4 85
     Phosphorus . . . . 1-50 gr.
     Morphine sulphate . . . . 1-12 gr.
     Zinc valerianate . . . . . 1 gr.
     Nervine, sedative. Dose, 1 to 2.

496. Phosphorus, Morphine and Zinc
     valerianate, "B" 1 00   4 85
     Phosphorus . . . . 1-25 gr.
     Morphine sulphate . . . . 1-12 gr.
     Zinc valerianate . . . . . 1 gr.
     Nervine, sedative. Dose, 1.

     Phosphorus and Nux Vomica. See
     Phosphorus Comp.

497. Phosphorus, Nux Vomica and
     Iron, "A" 30   1 35
     Phosphorus . . . . 1-100 gr.
     Extract Nux Vomica . . . . 1-2 gr.
     Ferric phosphate . . . . . 1-4 gr.
     Nervine, tonic. Dose, 1 to 2.

498. Phosphorus, Nux Vomica and
     Iron, "B" 40   1 85
     Phosphorus . . . . 1-33 gr.
     Extract Nux Vomica . . . . 1-3 gr.
     Reduced Iron . . . . 3 grs.
     Nervine, tonic. Dose, 1 to 2.

     Phosphorus, Nux Vomica, Iron
     and Quinine. See Quinine,
     Iron, Nux Vomica and Phos-
     phorus.

499. Phosphorus and Opium Com-
     pound, "A" 45   2 10
     Phosphorus . . . . 1-50 gr.
     Opium . . . . 1-4 gr.
     Digitalis . . . . 1-2 gr.
     Ipecac . . . . 1-4 gr.
     Nervine, tonic, sedative.
     Dose, 1 to 3.

500. Phosphorus and Opium Com-
     pound, "B" 45   2 10
     Phosphorus . . . . 1-33 gr.
     Opium . . . . 1-4 gr.
     Digitalis . . . . 1-2 gr.
     Ipecac . . . . 1-4 gr.
     Nervine, tonic, sedative.
     Dose, 1 to 2.

No. Per 100. Per 500.

Phosphorus and Quinine. 1 See
Quinine and Phosphorus.

Phosphorus, Quinine and Digitalis.
See Quinine and Digitalis
Comp.

Phosphorus, Quinine and Nux
Vomica. See Quinine, Nux
Vomica and Phosphorus.

501. Phosphorus and Strychnine "A" $ 45 $2 10
    Phosphorus . . . . 1-50 gr.
    Strychnine . . . . 1-60 gr.
    Nerve tonic. Dose, 1 to 2.

502. Phosphorus and Strychnine "B" 50 2 35
    Phosphorus . . . . 1-25 gr.
    Strychnine . . . . 1-30 gr.
    Nerve tonic. Dose, 1 to 2.

503. Phosphorus, Strychnine and Iron
    "A" 50 2 35
    Phosphorus . . . . 1-100 gr.
    Strychnine . . . . 1-60 gr.
    Vallet's mass . . . . 1 gr.
    Nervine, tonic. Dose, 1 to 2.

504. Phosphorus, Strychnine and Iron
    "B" 45 2 10
    Phosphorus . . . . 1-100 gr.
    Strychnine . . . . 1-100 gr.
    Ferrous sulphate . . . . . . 1 gr.
    Nervine, tonic. Dose, 1 to 2.

505. Phosphorus, Zinc and Strychnine. 50 2 35
    Phosphorus . . . . 1-70 gr.
    Zinc valerianate . . . . . . 3-4 gr.
    Strychnine . . . . 1-30 gr.
    Nervine, tonic. Dose, 1 to 2.

506. Phosphorus, Zinc and Valerian . 45 2 10
    Phosphorus . . . . 1-40 gr.
    Zinc sulphate . . . . 1 gr.
    Extract Valerian . . . . . . 2 grs.
    Sedative, nerve tonic. Dose, 1 to 2.

    Plummer's. See Antimony Comp.
    (69) 30 1 35

507. Podophyllin, 1-40 gr. 20 85
    Resin of Podophyllum, U. S. P.
    Cholagogue, cathartic.
    Dose, 1 to 3.

No.		Per 100.	Per 500.
508.	Podophyllin, 1-20 gr.	$ 20	$ 85
	Dose, 1 to 3.		
509.	Podophyllin, 1-10 gr.	20	85
	Dose, 1 to 3.		
510.	Podophyllin, 1-8 gr.	20	85
	Dose, 1 to 3.		
511.	Podophyllin, 1-6 gr.	20	85
	Dose, 1 to 3.		
512.	Podophyllin, 1-4 gr.	20	85
	Dose, 1 to 2.		
513.	Podophyllin, 1-2 gr.	25	1 10
	Dose, 1 to 2.		
514.	Podophyllin, 1 gr.	40	1 85
	Dose, 1.		
515.	Podophyllin and Belladonna	45	2 10

515. Podophyllin and Belladonna .. .. .45. 2 10
   Resin Podophyllum . . . . 1-4 gr.
   Extract Belladonna . . . . . 1-8 gr.
   Oleoresin Capsicum . . . . . 1-4 gr.
   Cholagogue, cathartic.
   Dose, 1 to 2.

516. Podophyllin, Belladonna and Calabar Bean . . . . 50 . 2 35
   Resin Podophyllum . . . . . 1-4 gr.
   Extract Belladonna . . . . . 1-4 gr.
   Extract Calabar Bean . . . . 1-4 gr.
   Cholagogue, cathartic.

517. Podophyllin and Blue Mass. . . . 30 1 35
   Resin Podophyllum . . . . 1-4 gr.
   Mercury mass . . . . 2 grs.
   Mercurial purgative. Dose, 1 to 2.

518. Podophyllin Compound 40 1 85
   Resin Podophyllum . . . . . 1-4 gr.
   Extract Mandrake . . . . . . 1 gr.
   Compound Extract Colocynth . 1 gr.
   Gentian . . . . 1-2 gr.
   Gambage . . . . 1-8 gr.
   Capsicum . . . . 1-16 gr.
   Cathartic. Dose, 1 to 3.

519. Podophyllin and Henbane 35 1 60
   Resin Podophyllum . . . . . 1-2 gr.
   Extract Hyoscyamus . . . . . 1-2 gr.
   Cholagogue, cathartic.
   Dose, 1 to 2.

No.		Per 100.	Per 500.

520. Podophyllin, Henbane and Ipecac $45 $2 10
  Resin Podophyllum . . . . . 2-3 gr.
  Extract Hyoscyamus. . . . . . 1 gr.
  Ipecac . . . . 1-3 gr.
  Cholagogue, cathartic.
  Dose, 1 to 2.

521. Podophyllin and Leptandrin . . 45 2 10
  Resin Podophyllum . . . . . 1-2 gr.
  Leptandrin . . . . 1 gr.
  Cholagogue, cathartic.
  Dose, 1 to 2.

522. Podophyllin and Macrotin . . . 45 2 10
  Resin Podophyllum . . . . . 1-4 gr.
  Macrotin . . . . 1-8 gr.
  Oleoresin Capsicum . . . . 1-8 gr.
  Antispasmodic, diuretic, cathartic.
  Dose, 1 to 2.

523. Podophyllin, Nux Vomica and
  Henbane 40 1 85
  Resin Podophyllum . . . . . 1-2 gr.
  Extract Nux Vomica . . . . 1-16 gr.
  Extract Hyoscyamus . . . . . 1-8 gr.
  Tonic, cholagogue, cathartic.

524. Potassium arsenate, 1-100 gr. . . 20 85
  Alterative. Dose, 1 to 3.

525. Potassium bromide, 1 gr. 45 2 10
  Sugar-coated only. See note 90.
  Nervous sedative, hypnotic.
  Dose, 1 to 2.

526. Potassium bromide, 5 grs. . . . 75 3 60
  Sugar-coated only. See note 90.
  Dose, 1 to 2.
  Potassium iodide, 2 grs. (665) . 55 2 60
  Sugar-coated only. See note 90.
  Alterative, tonic. Dose, 1 to 3.

527. Potassium iodide, 5 grs. 85 4 10
  Sugar-coated only. See note 90.
  Dose, 1 to 2.

528. Potassium permanganate, 1-8 gr. 30 1 35
  Sugar-coated only. See note 90.
  Antiseptic, emmenagogue.
  Dose, 1 to 3.

529. Potassium permanganate, 1-6 gr. 30 1 35
  Sugar-coated only. See note 90.
  Dose, 1 to 3.

No.		Per 100.	Per 500.
530.	Potassium permanganate, 1-4 gr. Sugar-coated only. See note 90. Dose, 1 to 3.	$130	$1 35
531.	Potassium permanganate, 1-2 gr. Sugar-coated only. See note 90. Dose, 1 to 3.	40	1 85
532.	Potassium permanganate, 1 gr. Sugar-coated only. See note 90. Dose, 1 to 2.	50	2 35
533.	Potassium permanganate, 2 grs. Sugar-coated only. See note 90. Dose, 1 to 2.	75	3 60
534.	Quassia, Extract, 1 gr. Bitter tonic. Dose, 1 to 3.	55	2 60
	Quevenne's Iron. See Iron, Reduced.		
535.	Quinine and Aloes Quinine sulphate . . . . 3-4 gr. Socotrine Aloes . . . . . 1-4 gr. Emmenagogue, tonic. Dose, 1 to 3.	60	2 85
536.	Quinine and Aloes Comp., "A," Hubbard's, stronger . . . . Quinine sulphate . . . . 1 1-2 grs. Purified Aloes . . . . . 2-3 gr. Piperine . . . . 2-3 gr. Strychnine sulphate . . . . 1-40 gr. Emmenagogue, tonic. Dose, 1 to 2.	90	4 35
537.	Quinine and Aloes Comp., "B," Hubbard's, milder . . . . . Quinine sulphate . . . . . . 1 gr. Purified Aloes . . . . . . 1-2 gr. Piperine . . . . 1-2 gr. Strychnine sulphate . . . . . 1-50 gr. Emmenagogue, tonic. Dose, 1 to 2.	75	3 60
538.	Quinine and Aloes Comp., "C" . . Quinine sulphate . . . . . . 1-2 gr. Socotrine Aloes . . . . . 1-2 gr. Extract Nux Vomica . . . 1-3 gr. Phosphorus . . . . 1-33 gr. Emmenagogue, tonic. Dose, 1 to 2.	60	2 85
539.	Quinine, Arsenic and Atropine, Dr. Whelan's . . . . Quinine sulphate . . . . . 1 1-2 grs. Arsenical solution, B. P. . . . 1 m. Extract Gentian . . . . . 1 2-3 grs. Solution Atropine sulph., B.P. 1-12 m. Antiperiodic, febrifuge. Dose, 1 to 3.	70	3 35

No.		Per 100.	Per 500.
540.	Quinine bisulphate, 1-4 gr.		
	Antiperiodic, febrifuge, tonic.		
	Dose, 1 to 3.		
541.	Quinine bisulphate, 1-2 gr.		
	Dose, 1 to 3.		
542.	Quinine bisulphate, 1 gr.		
	Dose, 1 to 3.		Prices
543.	Quinine bisulphate, 2 grs.		
	Dose, 1 to 3.		on
544.	Quinine bisulphate, 3 grs.		
	Dose, 1 to 2.		application.
545.	Quinine bisulphate, 4 grs.		
	Gelatin-coated only. See note 89.		
	Dose, 1 to 2.		
546.	Quinine bisulphate, 5 grs.		
	Gelatin-coated only. See note 89.		
	Dose, 1 to 2.		
	Quinine bromide. See Quinine hydrobrom.		
547.	Quinine and Capsicum	45	2 10
	Quinine sulphate . . . . 1 gr.		
	Capsicum . . . . . 1-2 gr.		
	Antiperiodic, febrifuge, tonic.		
	Dose, 1 to 3.		
548.	Quinine Compound	50	2 35
	Quinine sulphate . . . . 1 gr.		
	Arsenous acid . . . . 1-32 gr.		
	Reduced Iron . . . . 1 gr.		
	Antiperiodic, febrifuge, tonic.		
	Dose, 1 to 2.		
549.	Quinine Comp. and Dandelion	60	2 85
	Quinine bisulphate . . . . . 1 1-4 gr.		
	Arsenous acid . . . . 1-24 gr.		
	Dried Ferrous sulphate . . . . 2 grs.		
	Extract Dandelion . . . . 1 1-4 gr.		
	Antiperiodic, febrifuge, tonic.		
	Dose, 1 to 2.		
550.	Quinine Comp. and Strychnine	60	2 85
	Quinine sulphate . . . . 1 gr.		
	Arsenous acid . . . . 1-20 gr.		
	Reduced Iron . . . . 1 gr.		
	Strychnine . . . . 1-20 gr.		
	Antiperiodic, febrifuge, tonic.		
	Dose, 1 to 2.		

PILLS.

No. ·   Per 100. Per 500.

551. Quinine and Digitalis Comp. . . : $ 80 · . $3 85
  Quinine sulphate . .. : . . . . . 1-2 gr.
  Digitalis . . . . 1-2 gr.
  Ipecac . . . . , 1-4 gr.
  Opium . . . . 1-4 gr.
  Phosphorus . . . . 1-50 gr.
  Tonic, sedative. · Dose, 1 to 2.

552. Quinine hydrobrom., 1 gr. . . . 50   2 35
  Sedative, antiperiodic, febrifuge.
  Dose, 1 to 3.

553. Quinine hydrobrom., 2 grs. . . 75   3 60
  Dose, 1 to 3.

554. Quinine hydrobrom., 3 grs. . . 90   4 35
  Dose, 1 to 2.

555. Quinine and Iron, "A" 50   2 35
  Quinine sulphate . : . . . . . 1 gr.
  Reduced Iron . . . . 1 gr.
  Ferruginous tonic . Dose, 1 to 2. .

556. Quinine and Iron, "B" 45   2 10
  Quinine sulphate . . . . 1 gr.
  Vallet's mass . . . . 1 gr.
  Ferruginous tonic. Dose, 1 to 2.

557. Quinine, Iron, Nux Vomica and
  Phosphorus, "A" 50   2 35
  Quinine sulphate . . . . 1 gr.
  Reduced Iron . . . . 2 grs.
  Extract Nux Vomica . . . . 1-3 gr.
  Phosphorus . . . . 1-25 gr.
  Tonic, alterative, . Dose, 1 to 2. ·

558. Quinine, Iron, Nux Vomica and
  Phosphorus, "B" 50   2 35
  Quinine sulphate . . . . . . 1 gr.
  Vallet's mass . . . . 1 gr.
  Extract Nux Vomica . . . . 1-8 gr.
  Phosphorus . . . . . . 1-100 gr.
  Tonic, alterative. Dose, 1 to 2.

559. Quinine, Iron and Phosphorus,
  "A" 50   2 35
  Quinine sulphate . . . . . . 1 gr.
  Vallet's mass . . . . 1 gr.
  Phosphorus . . . . : . 1-100 gr.
  Tonic, alterative. Dose, 1 to 2.

No.   Per 100. Per 500.

560. Quinine, Iron and Phosphorus,
      "B" . . . . $.50 $2 35
      Quinine. . . . . 1-2 gr.
      Reduced Iron . . . . 3 grs.
      Phosphorus . . . . 1-50 gr.
      Tonic, alterative. Dose, 1 to 2.

561. Quinine, Iron, Phosphorus and
      Strychnine, "A"   50   2 35
      Quinine sulphate . . . . . 1-2 gr.
      Reduced Iron . . . . . . 1 1-5 grs.
      Phosphorus . . . . 1-50 gr.
      Strychnine . . . . 1-40 gr.
      Tonic, alterative. Dose, 1 to 2.

562. Quinine, Iron, Phosphorus and
      Strychnine, "B"   50   2 35
      Quinine sulphate . . . . . 1-2 gr.
      Reduced Iron . . . . 1 gr.
      Phosphorus . . . . 1-50 gr.
      Strychnine . . . . 1-60 gr.
      Tonic, alterative. Dose, 1 to 2.

563. Quinine, Iron and Strychnine .   50   2 35
      Quinine sulphate . . . 1 . 1 gr.
      Vallet's mass . . . . 2 grs.
      Strychnine sulphate . . . 1-60 gr.
      Tonic, alterative. Dose, 1 to 2.

564. Quinine, Iron and Strychnine
      phosphates, Easton's   65   3 10
      Quinine phosphate . . . . . 1 gr.
      Iron phosphate . . . . . . 1 gr.
      Strychnine phosphate . . . 1-60 gr.
      General tonic. Dose, 1 to 2.

565. Quinine and Iron valerianate .   80   3 85
      Quinine sulphate . . . . 1-2 gr.
      Iron valerianate . . . . . 1-2 gr.
      Tonic, alterative. Dose, 1 to 3.

566. Quinine, Iron and Zinc valerian-
      ates (Triple Valerianates) . .   95   4 60
      Quinine valerianate . . . . . 1 gr.
      Iron valerianate . . . . . . 1 gr.
      Zinc valerianate . . . . . . 1 gr.
      Tonic, antispasmodic. Dose, 1 to 3.

567. Quinine, Iron and Zinc valerian-
      ates, half-strength   50   2 35
      Quinine valerianate . . . . 1-2 gr.
      Iron valerianate . . . . . 1-2 gr.
      Zinc valerianate . . . . . 1-2 gr.
      Tonic, antispasmodic. Dose, 1 to 3.

PILLS.

No.		Per 100.	Per 500.
568.	Quinine, Nux Vomica and Phosphorus	$ 50	$2 35

Quinine sulphate . . . . 1 gr.
Extract Nux Vomica . . . 1-4 gr.
Phosphorus. . . . . . . 1-50 gr.
Tonic, nervine. Dose, 1 to 2.

569. Quinine and Phosphorus   50   2 35
Quinine sulphate . . . . . . 1 gr.
Phosphorus . . . . 1-50 gr.
Tonic, nervine. Dose, 1 to 2.

570. Quinine and Strychnine   50   2 35
Quinine sulphate . . . . . . 1 gr.
Strychnine . . . . . . 1-60 gr.
Tonic, nervine. Dose, 1 to 2.

571. Quinine sulphate, 1-4 gr. . . .
Antiperiodic, febrifuge, tonic.
Dose, 1 to 3.

572. Quinine sulphate, 1-2 gr. . . .
Dose, 1 to 3.

573. Quinine sulphate, 1 gr. . . .
Dose, 1 to 3.

574. Quinine sulphate, 2 grs. . . .
Dose, 1 to 3.

575. Quinine sulphate, 3 grs. . . .
Dose, 1 to 2.

576. Quinine sulphate, 4 grs. . . .
Gelatin-coated only. See note 89.
Dose, 1 to 2.

577. Quinine sulphate, 5 grs. . . .
Gelatin-coated only. See note 89.
Dose, 1 to 2.

Prices on application.

578. Quinine valerianate, 1-2 gr. . .   35   1 60
Antiperiodic, febrifuge, tonic.
Dose, 1 to 3.

579. Quinine valerianate, 1 gr. . . .   45   2 10
Dose, 1 to 3.

580. Quinine valerianate, 2 grs. . .   75   3 60
Dose, 1 to 3.

581. Quinine valerianate, 4 grs. . .   1 25   6 10
Dose, 1 to 2.

Reduced Iron. See Iron Reduced.

No....   Per 100. Per 500.

582. Rheumatic . . . . $ 55 $2.60
    Comp. Extract Colocynth . 1 1-2 grs.
    Extract Colchicum . . . . . . 1 gr.
    Mild Mercurous chloride . . . 1-8 gr.
    Extract Henbane . . . . 1-3 gr.
    Dose, 1 to 2.

583. Rhubarb, U. S. P.   40   1 85
    .20 Gm. Rhubarb . . . . . . 3 grs.
    .06 Gm. Soap . . . . 1 gr.
    Purgative, astringent, stomachic.
    Dose, 1 to 2.

584. Rhubarb and Blue Mass   40   1 85
    Rhubarb . . . . 1 3-4 grs.
    Mercury mass . . . . . . 1 3-4 grs.
    Sodium carbonate . . . . . 1-2 gr.
    Purgative. Dose, 1 to 2.

585. Rhubarb Comp., U. S. P.   40   1 85
    .13 Gm. Rhubarb . . . . . . 2 grs.
    .10 Gm. Purified Aloes . . 1 1-2 grs.
    .06 Gm. Myrrh . . . . 1 gr.
    .005 Cc. Oil Peppermint . 1-12 gr.
    Purgative. Dose, 1 to 2.

586. Rhubarb, Extract, 1 gr.   60   2 85
    Purgative, astringent, stomachic.
    Dose, 1 to 3.

587. Rhubarb and Iron   50   2 35
    Rhubarb . . . . 2 grs.
    Iron carbonate . . . . 1 gr.
    Purgative, astringent, stomachic.
    Dose, 1 to 3.

    Ricord's.   See Syphilitic "A"
    (637)   50   2 35
    Ricord's, Modified. See Syphilitic
    "B" (638) 1 00   4 85

588. Salicin, 1 gr.   40   1 85
    Antiseptic, antipyretic.
    Dose, 1 to 3.

589. Salicin, 2 grs.   55   2 60
    Dose, 1 to 3.

590. Salicin, 3 grs.   70   3 35
    Dose, 1 to 2.

591. Salicylic acid, 1 gr.   30   1 35
    Antiseptic, antipyretic, antirheu-
    matic. Dose, 1 to 3.

592. Salicylic acid, 2 grs.   40   1 85
    Dose, 1 to 3.

No.		Per 100.	Per 500.
593.	Salicylic acid, 3 grs. . . . . $	50	$2 35
	Dose, 1 to 3.		
594.	Salicylic acid, 5 grs.	65	3 10
	Gelatin-coated only. See note 89.		
	Dose, 1 to 2.		
595.	Salicylic acid and Morphine, "A"	90	4 35
	Salicylic acid . . . . . . 2 1-2 grs.		
	Morphine sulphate . . . . . . 1-12 gr.		
	Analgesic, antirheumatic.		
	Dose, 1 to 3.		
596.	Salicylic acid and Morphine, "B". 1	50	7 35
	Salicylic acid . . . . 5 grs.		
	Morphine sulphate . . . . . 1-8 gr.		
	Analgesic, antirheumatic.		
	Dose, 1 to 2.		
597.	Saline Chalybeate Tonic, Flint's.	40	1 85
	Sodium chloride. . . . . . . 3 grs.		
	Potassium chloride. . . . . 3-20. gr		
	Potassium sulphate . . . . 1-10 gr.		
	Potassium carbonate. . . . 1-20 gr.		
	Sodium carbonate. . . . . . 3-5 gr.		
	Magnesium carbonate . . . 1-20 gr.		
	Precip. Calcium phosphate . 1-2 gr.		
	Precip. Calcium carbonate . 1-20 gr.		
	Reduced Iron . . . . . . : 9-20 gr.		
	Vallet's mass . . . . 1-20 gr.		
	Dose, 1 to 2.		
598.	Salol, 2 1-2 grs.	70	3 35
	Intestinal antiseptic, antipyretic.		
	Dose, 1 to 2.		
599.	Salol, 5 grs. 1	00	4 85
	Dose, 1.		
600.	Sandalwood Comp.	70	3 35
	Oil Sandalwood . . . . 1 gr.		
	Extract Cubeb. . . . . 1 gr.		
	Copaiba . . . . 1 gr.		
	Antiblennorrhagic. Dose, 1 to 2.		
601.	Sanguinaria, Extract, 1-2 gr. . .	40	1 85
	Stimulant, narcotic, emmenagogue.		
	Dose, 1 to 2.		
602.	Santonin, 1-2 gr.	30	1 35
	Anthelmintic. Dose, 1 to 2.		
603.	Santonin, 1 gr.	40	1 85
	Dose, 1.		
604.	Santonin and Calomel	55	2 60
	Santonin. . . . . 1-2 gr.		
	Mild Mercurous chloride . . . 1-2 gr.		
	Anthelmintic. Dose, 1 to 2.		

No.		Per 100.	Per 500.
605.	Sedative, Warner's	$ .50	2 35
	Extract Sumbul . . . . 1-2 gr.		
	Extract Valerian . . . . . . 1-2 gr.		
	Extract Hyoscyamus. . . . . 1-2 gr.		
	Extract Cannabis Indica . . 1-10 gr.		
	Dose, 1 to 2.		
606.	Silver nitrate, 1-4 gr.	40	1 85
	Antispasmodic, tonic, astringent.		
	Dose, 1 to 2.		
607.	Soap Compound, B. P.	35	1 60
	Opium . . . . 1-2 gr.		
	Soap . . . . 2 grs.		
	Sedative, antacid. Dose, 1 to 2.		
608.	Sodium salicylate, 5 grs. . . .	70	3 35
	Antiseptic, antipyretic, antirheu-		
	matic. Dose, 1 to 2.		
609.	Squill Compound, U. S. P. '70 .	25	1 10
	Squill . . . . 1-2 gr.		
	Ginger . . . . 1 gr.		
	Ammoniac . . . . 1 gr.		
	Soap . . . . 1 1-2 grs.		
	Expectorant, diuretic. Dose, 1 to 2.		
610.	Stillingin, 1 gr.	35	1 60
	Alterative. Dose, 1 to 2.		
	Stork's. See Anti-Lacteous (62).	45	2 10
611.	Stramonium, Extract, 1-4 gr. . .	25	1 10
	Narcotic, diaphoretic.		
	Dose, 1 to 2.		
612.	Stramonium, Extract, 1-2 gr. . .	35	1 60
	Dose, 1 to 2.		
613.	Stramonium, Extract, 1 gr. . .	40	1 85
	Dose, 1 to 2.		
614.	Strophanthus, 1-20 gr.	50	2 35
	Cardiac tonic. Dose, 1 to 3.		
615.	Strophanthus, 1-4 gr. 1	00	4 85
	Dose, 1 to 2.		
616.	Strychnine, 1-100 gr.	20	85
	Nervine, tonic, stimulant.		
	Dose, 1 to 3.		
617.	Strychnine, 1-60 gr.	20	85
	Dose, 1 to 3.		
618.	Strychnine, 1-50 gr.	20	85
	Dose, 1 to 3.		
619.	Strychnine, 1-48 gr.	20	85
	Dose, 1 to 5.		

PILLS.

No.		Per 100.	Per 500.
620.	Strychnine, 1-40 gr.	$ 20	$ 85
	Dose, 1 to 2.		
621.	Strychnine, 1-32 gr.	20	85
	Dose, 1 to 2.		
622.	Strychnine, 1-30 gr.	20	85
	Dose, 1 to 2.		
623.	Strychnine, 1-20 gr.	20	85
	Dose, 1 to 2.		
624.	Strychnine, 1-16 gr.	20	85
	Dose, 1.		
625.	Strychnine, 1-12 gr.	20	85
	Dose, 1.		
626.	Strychnine arsenate, 1-200 gr.	20	85
	Nervine, tonic. Dose, 1 to 2.		
627.	Strychnine sulphate, 1-100 gr.	20	85
	Nervine, tonic, stimulant.		
	Dose, 1 to 2.		
628.	Strychnine sulphate, 1-50 gr.	20	85
	Dose, 1 to 2.		
629.	Strychnine sulphate, 1-30 gr.	20	85
	Dose, 1 to 2.		
630.	Sudorific	45	2 10
	Guaiac . . . . 1 1-2 grs.		
	Camphor . . . 1 1-2 grs.		
	Tartar Emetic . . . . 1-8 gr.		
	Extract Bittersweet, q. s.		
	Dose, 1 to 2.		
631.	Sulphonal, 2 1-2 grs.	1 80	8 85
	Hypnotic. Dose, 1 to 3.		
632.	Sulphonal, 5 grs.	3 60	17 85
	Dose, 1 to 2.		
633.	Sulphur iodide, 1-25 gr.	35	1 60
	Alterative, antiscorbutic.		
	Dose, 1 to 3.		
634.	Sulphur iodide, 1-10 gr.	35	1 60
	Dose, 1 to 2.		
635.	Sumbul Compound, "A"	1 09	4 85
	Extract Musk-root . . . . . . 1 gr.		
	Dried Ferrous sulphate . . . . 1 gr.		
	Asafetida . . . . 2 grs.		
	Arsenous acid . . . . 1-40 gr.		
	Tonic, antiperiodic, sedative.		
	Dose, 1 to 2.		

No.		Per 100.	Per 500.
636.	Sumbul Compound, "B"	$1 00	$4 85

    Extract Musk-root . . . . ¾ 1 gr.
    Asafetida . . . . 2 grs.
    Dried Ferrous sulphate . . . . 1 gr.
    Arsenous acid . . . . . . 1-100 gr.
    Tonic, antiperiodic, sedative.
    Dose, 1 to 2.

637. Syphilitic, "A," Ricord's 50   2 35

    Green Mercurous iodide . . 3-4 gr.
    Extract Opium . . . . . . 1-3 gr.
    Confection of Rose . . . . 1 1-2 grs.
    Dose, 1 to 2.

638. Syphilitic, "B," Ricord's Modi-
    fied 1 00   4 85

    Green Mercurous iodide . . 1-2 gr.
    Extract Conium fruit . . . 1 1-2 grs.
    Lactucarium . . . . 1-2 gr.
    Extract Opium . . 1) . 1. 1-10 gr.

Tanjore. See Arsenic, Fr. Codex
    (85) 20   85

Tartar Emetic. See Antimony
    and Potassium tartrate.

Taylor's. See Liver "D," (395). 35   1 60

639. Terpin hydrate, 2 grs. 40   1 85

    Stimulant, diuretic, anthelmintic.
    Dose, 1 to 2.

640. Terpin hydrate, 5 grs. 75   3 60

    Gelatin-coated only. See note 89.

641. Tonic, "A," Aiken's 55   2 60

    Quinine sulphate . . . . . . 1 gr.
    Reduced Iron . . . . . . 2-3 gr.
    Arsenous acid . . . . . . 1-50 gr.
    Strychnine . . . . 1-50 gr.
    Dose, 1 to 2.

642. Tonic, "B," Warner's 35   1 60

    Extract Gentian . . . . . . 1 gr.
    Extract Hops . . . . . . 1-2 gr.
    Extract Nux Vomica . . . 1-20 gr.
    Vallet's mass . . . . . . 1-4 gr.
    Resin Podophyllum . . . . 1-20 gr.
    Oleoresin Ginger . . . . 1-10 gr.
    Dose, 1 to 2.

Tonic, Waxham's. See Cinchon-
    ine Comp. (220) 40   1 85

PILLS.

No.		Per 100.	Per 500.

643. Tonic Hematic, Andrews' . . . . $ 75 . $3 60
    Quinine sulphate . . . . 1 gr.
    Reduced Iron . . . . 1 1-2 grs.
    Ipecac . . . . 1-8 gr.
    Arsenous acid . . . . 1-40 gr.
    Strychnine. . . . . 1-40 gr.
    Dose, 1 to 2.

644. Triplex, Common . . . . 40 . 1 85
    Aloes . . . . 2 grs.
    Mercury mass . . . . 1 gr.
    Resin Podophyllum . . . . . 1-4 gr.
    Cholagogue, cathartic.
    Dose, 1 to 3.

645. Triplex, Guy's Hospital . . . . 40 . 1 85
    Mercury mass . . . . 1 gr.
    Squill . . . . 1 gr.
    Digitalis . . . . 1 gr.
    Dose, 1 to 2.

646. Triplex Opt., Dr. Francis' . . . 40 . 1 85
    Aloes . . . . 1 1-5 grs.
    Resin Scammony . . . . 1 1-5 grs.
    Mercury mass . . . . . . 1 1-5 grs.
    Croton Oil . . . . 1-20 gr.
    Oil Caraway . . . . 1-4 gr.
    Tinct. Aloes and Myrrh . . . 1-3 gr.
    Cholagogue, cathartic.
    Dose, 1 to 2.

Triple Valerianates. See Quinine,
    Iron and Zinc Valerianates
    (566) . . . . 95 . 4 60

647. Valerian Extract, 2 grs. . . . . 55 . 2 60
    Antispasmodic, diaphoretic, seda-
      tive. Dose, 1 to 2.

Vallet's mass. See Ferrous car-
    bonate, Vallet's.

648. Veratrine, 1-60 gr. . . . . . . 25 . 1 10
    Powerful spinal and cardiac de-
      pressant. Dose, 1 to 2.

649. Veratrine, 1-32 gr. . . . . . . 30 . 1 35
    Dose, 1.

650. Veratrine, 1-12 gr. . . . . . . 35 . 1 60
    Dose, 1.

651. Veratrum Viride, Extract, 1-8 gr. 30 . 1 35
    Cardiac and spinal depressant.
    Dose, 1 to 3.

652. Veratrum Viride, Extract, 1-4 gr. 30 . 1 35
    Dose, 1 to 3.

No.		Per 100.	Per 500.
653.	Veratrum Viride, Extract, 1-2 gr. $	40	$1 85
	Dose, 1 to 2.		
	Viburnum Prunifolium. See Black Haw.		
	Wann's. See Bilious, "B," (123)	30	1 35
654.	Warburg's Tincture .	60	2 85
	(Each pill represents one fluidrachm of the tincture).		
	Antimalarial, tonic, alterative.		
	Dose, 1 to 2.		
655.	Warburg's Tincture, without Aloes .	60	2 85
	(Each pill represents one fluidrachm of the tincture, without aloes).		
	Antimalarial, tonic, alterative.		
	Dose, 1 to 2.		
	Warner's Antispasmodic. See Antispasmodic (80)	45	2 10
	Warner's Astringent. See Astringent (107)	35	1 60
	Warner's Laxative. See Laxative, "B" (384)	35	1 60
	Warner's Sedative. See Sedative (605).	50	2 35
	Warner's Tonic. See Tonic, "B," (642)	35	1 60
	Watson's Anti-Chill. See Chinoldin Comp., "B" (194).	40	1 85
	Waxham's Liver. See Liver, "E" (396)	30	1 35
	Waxham's Tonic. See Cinchonine Comp. (220)	40	1 85
	Whelan's. See Quinine, Arsenic and Atropine (539)	70	3 35
656.	Zinc phosphide, 1-10 gr.	30	1 35
	Nerve stimulant, tonic.		
	Dose, 1 to 3.		
657.	Zinc phosphide, 1-6 gr.	35	1 60
	Dose, 1 to 3.		
658.	Zinc phosphide, 1-4 gr.	40	1 85
	Dose, 1 to 2.		

No.	Per 100.	Per 500.
659. Zinc phosphide, 1-2 gr. ... $ 50.		$2 35
Dose, 1 to 2.		
660. Zinc phosphide Comp., "A" ... 65		3 10
Zinc phosphide . . . . . 1–8 gr.		
Extract Nux Vomica . . . 1–8 gr.		
Extract Cannabis Indica . . 1–8 gr.		
Nervine, tonic. Dose, 1 to 2.		
661. Zinc phosphide Comp., "B" . . 35		1 60
Zinc phosphide . . . . . 1–10 gr.		
Extract Nux Vomica . . . 1–4 gr.		
Nervine, tonic. Dose, 1 to 2.		
662. Zinc phosphide Comp., "C" . . 60		2 85
Zinc phosphide . . . . . 1–10 gr.		
Extract Nux Vomica . . . 1–4 gr.		
Quinine sulphate . . . . 1 gr.		
Nervine, tonic. Dose, 1 to 2.		
663. Zinc valerianate, 1 gr. . . . 40		1 85
Nervine, antispasmodic.		
Dose, 1 to 3.		
664. Zinc valerianate, 2 grs. . . . 50		2 35
Dose, 1 to 2.		
665. Potassium iodide, 2 grs. . . . 55		2 60

EFFERVESCENT GRANULAR PREPARATIONS.

Each drachm contains the specified quantity of medicinal ingredients.

	Per ℔. bottle.
Caffeine, citrated, U. S. P., 1 1-5 grs. . . .	$1 50
Caffeine hydrobromate, 1 gr.	1 50
Caffeine hydrobromate Comp.	1 50
Caffeine hydrobromate . . . . 1 gr.	
Celery . . . . 1 gr.	
Acetanilid . . . . . . 2 grs.	
Sodium bromide . . . . . 5 grs.	
Congress	1 50
Lithium benzoate, 4 grs.	3 50
Lithium citrate, U. S. P., 8 grs. . . .	3 50
Lithium salicylate, 4 grs.	3 50
Magnesium citrate	1 50
Potassium bromide, 10 grs.	1 50
Seltzer	1 50
Vichy.	1 50

TINTED GRANULES.

The principal difference between pills and tinted granules is that the latter have a pink coating and contain much smaller amounts of the active ingredients. For this reason they are well suited for administration to children and in cases where it is desired to exhibit the medicine in minute, frequently repeated doses.

No.		Per 100.	Per 500.
1	Aconite root, 1-20 gr.	$ 20	$ 85
2	Aloin, 1-10 gr.	20	85
3	Alum, 1-10 gr.	20	85
4	Ammonium chloride, 1-10 gr.	20	85
5	Anodyne	20	85
	Camphor. . . . . 1–8 gr.		
	Extract Hyoscyamus. . . . . . 1–8 gr.		
	Morphine acetate . . . . . . 1–160 gr.		
	Oleoresin Capsicum . . . . 1–160 gr.		
6	Anti-Chill	20	85
	Chinoidine . . . . 1–16 gr.		
	Iron ferrocyanide . . . . . . 1–8 gr.		
	Arsenous acid . . . . 1–320 gr.		
	Oleoresin Black Pepper . . . . 1–16 gr.		
7	Anti-Constipation	20	85
	Resin Podophyllum . . . . . 1-4 gr.		
	Capsicum . . . . 1–16 gr.		
	Extract Nux Vomica . . . . . 1–16 gr.		
	Extract Belladonna . . . . . 1–40 gr.		
	Extract Hyoscyamus . . . . 1–16 gr.		
8	Arnica flowers, 1-5 gr.	20	85
9	Arsenous acid, 1-100 gr.	20	85
10	Arsenous acid, 1-60 gr.	20	85
11	Arsenous acid, 1-50 gr.	20	85
12	Arsenous acid, 1-30 gr.	20	85
13	Arsenous iodide, 1-100 gr.	20	85
14	Belladonna leaves, 1-20 gr.	20	85
15	Calomel, 1-20 gr.	20	85
	"Mild Mercurous chloride."		
16	Calomel, 1-8 gr.	20	85
17	Calomel, 1-5 gr.	20	85
18	Camphor, 1-20 gr.	20	85

No.		Per 100.	Per 500.
19	Camphor and Opium	$ 20	$ 85
	Camphor . . . . 1-6 gr.		
	Powd. Opium . . . . 1-12 gr.		
20	Camphor, Opium and Tannin . . .	20	85
	Camphor . . . . 1-16 gr.		
	Opium . . . . 1-64 gr.		
	Tannin . . . . 1-8 gr.		
21	Cantharides, 1-50 gr.	20	85
22	Cathartic Comp., Improved	20	85
	Compound Ext. Colocynth . . 1-12 gr.		
	Extract Jalap . . . . 1-24 gr.		
	Resin Podophyllum . . . . . 1-48 gr.		
	Leptandrin . . . . 1-48 gr.		
	Extract Hyoscyamus . . . . . 1-48 gr.		
	Extract Gentian. . . . . 1-24 gr.		
	Oil Peppermint, q. s.		
23	Cathartic Comp., U. S. P. mass . .	20	85
	Compound Extract Colocynth . 1-11 gr.		
	Extract Jalap . . . . 1-28 gr.		
	Mild Mercurous chloride . . . . 1-14 gr.		
	Gamboge . . . . 1-56 gr.		
24	Corrosive sublimate, 1-100 gr. . . .	20	85
	Corrosive Mercuric chloride. .		
25	Corrosive sublimate, 1-40 gr. . . .	20	85
26	Croton Oil, 1-50 gr.	20	85
27	Digitalis, 1-20 gr.	20	85
28	Dover's Powder, 1-4 gr.	20	85
	Powder of Ipecac and Opium.		
29	Ergotin, Bonjean's, 1-10 gr. . . .	20	85
30	Gamboge, 1-32 gr.	20	85
31	Gelsemium, 1-50 gr.	20	85
32	Hydrastin, 1-20 gr.	20	85
33	Hyoscyamus, Extract, 1-8 gr. . . .	20	85
34	Iodoform, 1-10 gr.	20	85
35	Ipecac, 1-50 gr.	20	85
36	Iron, Reduced, 1-10 gr.	20	85
37	Jalap resin, 1-16 gr.	20	85
38	Leptandrin, 1-16 gr.	20	85
39	Mercurous iodide, yellow, 1-20 gr. .	20	85
40	Mercury with Chalk, 1-10 gr. . . .	20	85
41	Morphine sulphate, 1-50 gr. . . .	20	85

No.	POSIMETRIC GRANULES.	Per 100.	Per 500.
42	Neuralgic, Dr. Gross'	$ 20	$ 85
	Quinine sulphate . . . . 1-5 gr.		
	Morphine sulphate . . . . . . 1-200 gr.		
	Arsenous acid . . . . 1-200 gr.		
	Extract Aconite leaves' . . . . . 1-20 gr.		
	Strychnine . . . . 1-300 gr.		
43	Nux Vomica, 1-50 gr.	20	85
44	Opium, 1-40 gr.	20	85
45	Peppermint Oil, 1-128 gr.	20	85
46	Phosphorus, 1-200 gr.	20	85
47	Piperin, 1-20 gr.	20	85
48	Podophyllum resin, 1-40 gr.	20	85
49	Podophyllum resin, 1-4 gr.	20	85
50	Potassium arsenate, 1-100 gr. . . .	20	85
51	Potassium bromide, 1-5 gr.	20	85
52	Potassium nitrate, 1-10 gr.	20	85
53	Quinine sulphate, 1-10 gr.	20	85
54	Quinine, Iron and Strychnine . . .	20	85
	Quinine sulphate . . . . . . 1-16 gr.		
	Vallet's mass . . . . 1-8 gr.		
	Strychnine sulphate . . . . . 1-960 gr.		
55	Salicylic acid, 1-10 gr.	20	85
56	Santonin, 1-10 gr.	20	85
57	Strychnine, 1-100 gr.	20	85
58	Strychnine, 1-60 gr.	20	85
59	Tannic acid, 1-20 gr.	20	85
60	Tartaric acid, 1-10 gr.	20	85

OLEORESINS.

	Per oz.
Black Pepper	$1 10
Capsicum	70
Cubeb	75
Ginger	1 50
Male-fern	95

DOSIMETRIC GRANULES.

There is much to be said in favor of administering the isolated active principles of drugs instead of the more bulky and less palatable forms, such as tinctures, extracts, powders, etc. In many cases, and particularly where it has been shown that the activity of a drug is due to the presence of a definite chemical compound, the results are much more prompt and satisfactory, and are obtained with a certainty that has won for this system the name, "Positive Medication." The fact that there are many drugs whose medicinal effects cannot be traced to any proximate principle does not in the least affect the rationale of administering true active principles, and this is all that is claimed for positive medication by its advocates. The alkaloids and other organic drugs used in the preparation of the following named granules are of the best grades obtainable in the markets. In addition to the organic drugs we have included in the following list some of the more active inorganic remedial agents.

No.		Per 100.	Per 500.
1	Atropine, 1-256 gr.	$ 30	$1 35
2	Atropine, 1-128 gr.	30	1 35
3	Atropine sulphate, 1-256 gr.	30	1 35
4	Atropine sulphate, 1-128 gr.	30	1 35
5	Atropine valerianate, 1-256 gr.	30	1 35
6	Atropine valerianate, 1-128 gr.	40	1 85
7	Benzoic acid, 5-32 gr.	20	85
8	Bismuth subnitrate, 5-32 gr.	20	85
9	Brucine, 1-128 gr.	20	85
10	Bryonin, 1-64 gr.	30	1 35
11	Cadmium sulphate, 5-32 gr.	20	85
12	Caffeine arsenate, 1-64 gr.	30	1 35
13	Caffeine, citrated, 5-32 gr.	30	1 35
14	Caffeine valerianate, 5-32 gr.	30	1 35
15	Calcium hypophosphite, 5-32 gr.	20	85

No.		Per 100.	Per 500.
16	Calcium sulphide, 5-32 gr. . . .	$ 20	$ 85
17	Camphor, monobromated, 5-32 gr.	20	85
18	Cocaine, 1-64 gr.	30	1 35
19	Cocaine hydrochlor., 1-64 gr. . .	30	1 35
20	Cocaine hydrochlor., 5-32 gr. . .	50	2 35
21	Codeine, 5-32 gr.	75	3 60
22	Codeine, 1-4 gr.	90	4 35
23	Colchicine, 1-128 gr.	30	1 35
24	Colchicine, 1-64 gr.	40	1 85
25	Colocynthin, 1-64 gr.	30	1 35
26	Coniine hydrochlor., 1-64 gr. . .	30	1 35
27	Convallamarin, 1-128 gr. . . .	50	2 35
28	Croton oil, 5-32 gr.	30	1 35
29	Cubebin, 1-64 gr.	30	1 35
30	Curare, 1-64 gr.	50	2 35
31	Daturine, 1-256 gr.	40	1 85
32	Daturine, 1-128 gr.	60	2 85
33	Diastase, 5-32 gr.	20	85
34	Digitalin, 1-128 gr.	30	1 35
35	Digitalin, 1-64 gr.	40	1 85
36	Duboisine sulphate, 1-128 gr. . .	50	2 35
37	Elaterin, 1-64 gr.	50	2 35
38	Emetine, 1-64 gr.	50	2 35
39	Ergotin, 5-32 gr.	20	85
40	Ferrous iodide, 5-32 gr. . . .	20	85
41	Gelsemine, 1-64 gr.	50	2 35
42	Homatropine, 1-64 gr. . . .	50	2 35
43	Hydrastine, 1-64 gr.	30	1 35
44	Hydrastine hydrochlor., 5-32 gr. .	40	1 85
45	Hyoscine hydrobrom., 1-256 gr. .	40	1 85
46	Iodoform, 5-32 gr.	20	85
47	Iron arsenate, 1-64 gr.	20	85
48	Jalap resin, 1-64 gr.	20	85
49	Kosin, 1-64 gr.	30	1 35
50	Lithium benzoate, 5-32 gr. . .	45	2 10
51	Lobeline, 1-64 gr.	90	4 35
52	Manganese arsenate, 1-64 gr. .	120	85

No.		Per 100.	Per 500.
53	Mercuric chloride, corros., 1-64 gr.	$ 20	$ 85
54	Mercuric iodide, red, 1-64 gr. . .	20	85
55	Mercurous chloride, mild, 5-32 gr.	20	85
56	Mercurous iodide, yellow, 1-64 gr.	20	85
57	Morphine hydrobrom., 1-64 gr. .	30	1 35
58	Morphine hydrochlor., 1-64 gr. .	30	1 35
59	Morphine sulphate, 1-64 gr. . . .	20	85
60	Morphine sulphate, 5-32 gr. . . .	30	1 35
61	Muscarine sulphate, 1-64 gr. . . .	1 60	7 85
62	Napelline, 1-128 gr.	60	2 85
63	Narceine, 1-64 gr.	90	4 35
64	Pancreatin, 5-32 gr.	20	85
65	Pelletierine, 1-64 gr.	1 00	4 85
66	Pepsin, 5-32 gr.	20	85
67	Physostigmine salicylate, 1-128 gr.	80	3 85
68	Picrotoxin, 1-128 gr.	40	1 85
69	Pilocarpine hydrochlor., 1-64 gr. .	45	2 10
70	Pilocarpine nitrate, 1-64 gr. . . .	45	2 10
71	Piperine, 1-64 gr.	20	85
72	Podophyllotoxin, 1-64 gr.	30	1 35
73	Quassin, amorphous, 1-5 gr. . . .	35	1 60
74	Quassin, cryst., 5-32 gr.	1 10	5 35
75	Quinine arsenate, 1-64 gr.	20	85
76	Quinine hydrobrom., 5-32 gr. . .	50	2 35
77	Quinine salicylate, 5-32 gr. . . .	30	1 35
78	Quinine sulphate, 5-32 gr.	20	85
79	Quinine valerianate, 5-32 gr. . . .	30	1 35
80	Salicylic acid, 5-32 gr.	20	85
81	Sanguinarine sulphate, 1-64 gr. . .	30	1 35
82	Santonin, 5-32 gr.	20	85
83	Scillitin, 1-256 gr.	25	1 10
84	Scillitoxin, 1-256 gr.	45	2 10
85	Sodium benzoate, 5-32 gr.	20	85
86	Sodium hypophosphite, 5-32 gr. .	20	85
87	Sodium salicylate, 5-32 gr. . . .	20	85
88	Strychnine arsenate, 1-128 gr. . .	20	85
89	Strychnine sulphate, 1-256 gr. . .	20	85

No.	PODERMIC TABLETS.	Per 100.	Per 500.
90	Strychnine sulphate, 1-128 gr. . . .	$ 20	$ 85
91	Sulphur iodide, 5-32 gr. , ...	25	I 10
92	Tannic acid, 5-32 gr. . . .. . , . .	20	85
93	Uranium nitrate, 5-32 gr.	20	85
94	Veratrine, 1-128 gr. . . . ". . .	25	I 10
95	Zinc phosphide, 1-128 grs.	20	85
96	Zinc phosphide, 1-64 gr.	20	85

SPIRITS.

Per pint.

Ammonia, U. S. P.	$ 75
Ammonia, Aromatic, U. S. P.	75
Anise, U. S. P. . . . , ...	90
Aromatic, N. F.	I 20
Aromatic, Stearns' . . , ...	I 25
Bay, U. S. P.	65
Camphor, U. S. P.	85
Chloroform, U. S. P.	I 00
Cinnamon, U. S. P.	90
Ether, U. S. P.	85
Ether Comp., U. S. P.	I 25
Hoffman's Anodyne.	
Juniper, U. S. P.	60
Juniper Comp., U. S. P.	60
Lavender, U. S. P.	75
Lavender Comp.	I 00
Comp. Tincture Lavender.	
Lemon, U. S. P.	85
Nitrous Ether, U. S. P.	85
Nitrous Ether, Conc.	I 50
Eight times as strong as the U. S. P. Spirit.	
Nutmeg, U. S. P.	75
Orange, U. S. P.	90
Orange Comp.	I 00
Peppermint, U. S. P.	I 25
Spearmint, U. S. P.	I 25
Wintergreen, U. S. P.	75

HYPODERMIC TABLETS.

Our Hypodermic Tablets are readily soluble and strictly permanent, and are prepared from the best obtainable materials. Unless otherwise directed on orders, we will send them in tubes of 25 each. If bottles of 100 are wanted, it will be necessary to specify "in 100's" or "in bottles of 100" on the order. Special quotations made on private formula work.

No.		Per tube of 25	Per 100 in tubes of 25	Per bottle of 100
1	Aconitine, cryst., 1-240 gr. . .	$ 16	$ 60	$ 50
2	Aconitine, cryst., 1-120 gr. . .	.22	80	70
3	Apomorphine hydrochlorate, 1-12 gr.	20	75	65
4	Apomorphine hydrochlorate, 1-6 gr.	32	1 20	1 10
5	Atropine sulphate, 1-120 gr. . .	12	40	30
6	Atropine sulphate, 1-60 gr. . . .	12	40	30
7	Atropine and Digitalin. . . .	27	1 00	90
	Atropine sulphate . . . 1-120 gr.			
	Digitalin. . . . . . . gr.			
8	Cocaine hydrochlorate, 1-8 gr. .	16	60	50
9	Cocaine hydrochlorate, 1-4 gr.	22	80	70
10	Codeine sulphate, 1-6 gr. . .	20	70	60
11	Colchicine, 1-50 gr.	27	1 00	90
12	Coniine hydrochlor., 1-120 gr.	12	40	30
13	Coniine hydrochlor., 1-60 gr. .	14	50	40
14	Digitalin, 1-60 gr.	14	50	40
15	Ergotin, 1-24 gr.	12	40	30
16	Ergotin, 1-12 gr.	14	50	40
	Eserine. See Physostigmine.			
17	Gelsemine hydrochlor. 1-120 gr.	35	1 35	1 25
18	Homatropine, 1-120 gr. . . .	45	1 75	1 65
19	Hyoscine hydrobrom., 1-120 gr.	18	65	55
20	Hyoscyamine, 1-120 gr. . . .	14	50	40
21	Hyoscyamine, 1-60 gr.	16	60	50

No.	ED TABL	Per tube of 25	Per 100 in tubes of 25	Per bottle of 100
22	Morphine sulphate, 1-12 gr.	$ 12	$ 40	$ 30
23	Morphine sulphate, 1-8 gr.	12	40	30
24	Morphine sulphate, 1-6 gr.	13	45	35
25	Morphine sulphate, 1-4 gr.	14	50	40
26	Morphine sulphate, 1-3 gr.	15	55	45
27	Morphine sulphate, 1-2 gr.	20	75	65
28	Morphine and Atropine, "A"	18	65	55
	Morphine sulphate. 1-3 gr.			
	Atropine sulphate 1-120 gr.			
29	Morphine and Atropine, "B"	15	55	45
	Morphine sulphate. 1-6 gr.			
	Atropine sulphate 1-180 gr.			
30	Morphine and Atropine, "C"	15	55	45
	Morphine sulphate 1-12 gr.			
	Atropine sulphate 1-240 gr.			
31	Physostigmine salicyl., 1-120 gr.	15	55	45
32	Physostigmine salicyl., 1-60 gr.	20	75	65
33	Pilocarpine hydrochlor.,1-18 gr.	20	70	60
34	Pilocarpine hydrochlor.,1-12 gr.	30	1 10	1 00
35	Pilocarpine hydrochlor., 1-6 gr.	40	1 50	1 40
36	Strychnine sulphate, 1-120 gr.	12	40	30
37	Strychnine sulphate, 1-60 gr.	12	40	30
38	Veratrine, 1-30 gr.	25	90	80

GLYCERITES.

	Per pint.
Boroglyceride, 50 %	$1 50
Pepsin and Wafer Ash	90

Each fluidounce represents—
Saccharated Pepsin . 80 grs.
Wafer Ash . 30 grs.

Tar	65
Yerba Santa Compound	80

Each fluidounce represents—
Yerba Santa . 90 grs.
Licorice . 90 grs.
Wild Cherry . 30 grs.
Salicylic acid . 2 grs.
Potassium bromide 1-2 grs.
Grindelia . 30 grs.
Tar . 3-4 grs.

COMPRESSED TABLETS.

The use of compressed tablets has grown to such an extent that comment upon them seems superfluous. We list the most staple of these goods at prices as low as is consistent with the best grade of materials and skilled manufacture. It is a not uncommon ruse among some manufacturers who quote extremely low prices to either omit or greatly reduce the quantities of the more expensive ingredients in combinations, while retaining the names established by common custom and long usage. This practice is wholly reprehensible and should be discouraged. Therefore we request comparisons of our formulæ with those of other houses whenever comparisons of prices are made.

No.		Per 1,000.	Per 500.	Per 100.
1	Acetanilid, 2 grs.	$ 83	$ 47	$ 15
2	Acetanilid, 3 grs.	1 00	55	15
3	Acetanilid, 4 grs.	1 25	67	19
4	Acetanilid, 5 grs.	1 34	70	20
5	Aloes, U. S. P.	1 85	1 00	25
	.13 Gm. Aloes. . . . . . . 2 grs.			
	.13 Gm. Soap . . . . . . 2 grs.			
6	Aloes and Mastic, U. S. P. .	1 85	1 00	25
	.13 Gm. Purif. Aloes. . . . 2 grs.			
	.04 Gm. Mastic . . . . . 3-5 gr.			
	.03 Gm. Red Rose . . . 1-2 gr.			
7	Aloin, Strychnine and Belladonna	1 50	80	20
	Aloin . . . . 1-5 gr.			
	Strychnine . . . . 1-60 gr.			
	Ext. Belladonna . . . . . 1-8 gr.			
8	Ammonium bromide, 5 grs. .	1 60	83	22
9	Ammonium bromide, 10 grs.	1 90	1 00	25
10	Ammonium chloride, 3 grs. .	60	34	15
11	Ammonium chloride, 5 grs.	68	38	15
	Per pound jar, $0 70.			
12	Ammonium chloride, 10 grs.	90	50	15

No.		Per 1,000.	Per 500.	Per 100.
13	Anti-Bilious "A," Cook's . .	$1 85	$1 00	$ 25
	Aloes . . . . 1 gr.			
	Rhubarb. . . . . 1 gr.			
	Mild Mercurous chloride . 1-2 gr.			
	Soap. . . . . 1-2 gr.			
14	Anti-Bilious "B," Vegetable.	2 50	1 30	30
	Comp. Ext. Colocynth, 2 1-2 grs.			
	Resin Podophyllum . . . 1-4 gr.			
15	Anti-Constipation, Brundage .	1 90	1 00	25
	Ext. Nux Vomica . . . . 1-4 gr.			
	Ext. Hyoscyamus . . . . 1-4 gr.			
	Ext. Belladonna. . . . . 1-10 gr.			
	Capsicum . . . . . 1-4 gr.			
	Resin Podophyllum . . . . 1-10 gr.			
16	Antimony Comp., U. S. P., Plummer's	1 30	70	18
	.04 Gm. Sulph. Antimony. 3-5 gr.			
	.04 Gm. Mild Merc. chlor. 3-5 gr.			
	.08 Gm. Guaiac . . . . 1 1-4 grs.			
17	Anti-Pain	3 35	1 85	45
	Acetanilid. . . . . 4 grs.			
	Tartaric acid . . . . . . 1-4 gr.			
	Caffeine . . . . 1-10 gr.			
	Sodium bicarbonate . . 13-20 gr.			
18	Antipyrine, 3 grs.	. .	12 75	2 60
19	Antipyrine, 5 grs.	. .	19 60	4 00
20	Antipyrine, 10 grs.	. .	37 75	7 65
21	Antiseptic, "A," Sealer's . .	2 00	1 05	25
	Per pound jar, $1 50.			
	Sodium bicarbonate . . 3 3-4 grs.			
	Sodium borate. . . . . 3 3-4 grs.			
	Sodium benzoate . . . . 5-32 gr.			
	Sodium salicylate . . . . 5-32 gr.			
	Eucalyptol . . . . . . 5-64 gr.			
	Thymol . . . . 5-64 gr.			
	Menthol. . . . . 5-128 gr.			
	Oil Wintergreen . . . . 1-20 m.			
	See note 6.			
22	Antiseptic, "B," pink. . . .	4 00	2 05	67
	Per bottle of 25, $0 25.			
	Corros. Mercuric chlor. 7 3-10 grs.			
	Citric acid . . . . . . 3 4-5 grs.			
	For external use only.			
23	Aperient	2 50	1 30	30
	Ext. Nux Vomica . . . 1-3 gr.			
	Ext. Hyoscyamus . . . . 1-2 gr.			
	Comp. Ext. Colocynth . . 2 grs.			

No.		Per 1,000.	Per 500.	Per 100.
24	Borax, 5 grs. . . . . Per pound jar, $0 70.	$ 60	$ 34	$ 15
25	Boric acid, 5 grs.	1 00	55	15
26	Calomel and Soda . . . . . Mild Merc. chloride . . 2 1-2 grs. Sodium bicarbonate . . 2 1-2 grs.	1 25	67	19
27	Cathartic Comp., U. S. P. . .08 Gm. Co. Ext. Coloc. 1 1-4 grs. .06 Gm. Mild Merc. chlor. 1 gr. .03 Gm. Ext. Jalap . . . 1-2 gr. .015 Gm. Gamboge. . . . 1-4 gr.	2 00	1 05	25
28	Cathartic Comp., Improved . Comp. Ext. Colocynth . . . 1 gr. Ext. Hyoscyamus . . . 1-4 gr. Resin Podophyllum . . . 1-4 gr. Ext. Jalap . . . . 1-2 gr. Ext. Gentian. . . . . . 1-2 gr. Leptandrin. . . . . . . 1-4 gr. Oil Peppermint. . . . . . q. s.	2 00	1 05	25
29	Digestive . . . . Lactinated Pepsin. . . . 5 grs. See note 71 a.	3 50	1 80	40
30	Dover's Powder, 2 grs. . . . Powder of Ipecac and Opium.	1 10	60	17
31	Dover's Powder, 3 grs. . . .	1 40	75	20
32	Dover's Powder, 5 grs. . . .	1 85	1 00	25
33	Dyspepsia, Finch's . . . . . Rhubarb. . . . . 1-2 gr. Cubeb . . . . 1-2 gr. Calcined Magnesia. . . . 1 gr.	1 50	80	20
34	Ferrous carb., Vallet's, 3 grs.	1 50	80	20
	Kermes Mineral, 1-4 gr. . . See Sulphurated Antimony (94).	2 50	1 30	30
35	Little Liver . . . . Aloin . . . . 1-10 gr. Resin Jalap. . . . . . . 1-10 gr. Resin Podophyllum . . . 1-5 gr. Ext. Hyoscyamus . . . . 1-20 gr. Ext. Nux Vomica . . . 1-20 gr. Oleoresin Capsicum . . . 1-20 gr.	1 50	80	20

No.		Per 1,000	Per 500.	Per 100.
36	Mercurial (Blue Mass), 5 grs.	$2 00	$1 05	$ 25
	Mercury Mass.	but		
37	Morphine sulphate, 1-32 gr.			
38	Morphine sulphate, 1-20 gr.			
39	Morphine sulphate, 1-12 gr.			
40	Morphine sulphate, 1-10 gr.			
41	Morphine sulphate, 1-8 gr.			
42	Morphine sulphate, 1-6 gr.			
43	Morphine sulphate, 1-4 gr.			
44	Opium, 1 gr.	3 00	1 55	35
45	Pepsin, Saccharated, 5 grs.	1 60	85	22
46	Pepsin and Bismuth . . .	4 25	2 30	50
	Saccharated Pepsin . . 3 grs.			
	Bismuth subnitrate . . . 5 grs.			
47	Pepsin, Bismuth, and Strych-			
	-nine	3 00	1 55	35
	Saccharated Pepsin . . . 3 grs.			
	Bismuth subnitrate . . . 2 grs.			
	Strychnine . . . . 1-60 gr.			
48	Pepsin and Calcium lactophos-			
	phate . . . . . 1 & .	2 50	1 30	30
	Saccharated Pepsin . . . 3 grs.			
	Calcium lactophosphate . . 2 grs.			
49	Pepsin Compound	4 75	2 50	55
	Pepsin. . . . . . 1 gr.			
	Pancreatin. . . . . 1 gr.			
	Calcium lactophosphate . . 1 gr.			
	Lactic acid. . . . . . q. s.			
50	Phenacetine, 1 gr.	6 20	3 15	67
51	Phenacetine, 2 grs. . . .	11 85	5 95	1 23
52	Phenacetine, 3 grs. . . .	17 25	8 65	1 78
53	Podophyllin, 1-8 gr. . .	1 05	60	17
	Resin of Podophyllum.			
54	Podophyllin, 1-4 gr. . .	1 35	75	20
55	Podophyllin, 1-2 gr. . .	1 75	95	25
56	Potassium bicarbonate, 5 grs.	83	47	15
	Per pound jar, $1 00.			
57	Potassium bicarbonate, 10 grs.	1 10	60	17
58	Potassium bromide, 5 grs. .	1 20	64	18
59	Potassium bromide, 10 grs. .	1 90	1 00	25
60	Potassium chlorate, 5 grs.	67	40	15
	Per pound jar, $0 70.			

No.		Per 1,000.	Per 500.	Per 100.
61	Potassium chlorate and Borax.	$ 67	$ 40	$ 15
	Per pound jar, $0 80.			
	Potassium chlorate . . 2 1-2 grs.			
	Sodium borate. . . . 2 1-2 grs.			
62	Potassium chlorate and [1]Borax Comp.	1 50	80	20
	Potassium chlorate . . 2 1-2 grs.			
	Sodium borate . . . . . 2 1-2 grs.			
	Cocaine hydrochlor. . . . 1-40 gr.			
63	Potassium iodide, 5 grs. . . .	6 90	3 50	75
64	Potassium permangan., 1-2 gr.	50	30	15
65	Potassium permanganate, 1 gr.	50	30	15
66	Potassium permanganate, 2 grs.	60	34	15
67	Potassium permanganate, 3 grs.	67	37	15
68	Potassium permanganate, 5 grs.	83	47	15
69	Quinine sulphate, 1 gr. . . .			
70	Quinine sulphate, 2 grs. . . .			
71	Quinine sulphate, 3 grs. . .	Prices on application.		
72	Quinine sulphate, 5 grs. . .			
73	Quinine tannate, 1 gr. (with chocolate)			
74	Quinine tannate, 2 1-2 grs. (with chocolate)			
75	Rhubarb, U. S. P.	1 50	80	20
	.20 Gm. Rhubarb . . . . . 3 grs.			
	.06 Gm. Soap . . . . 1 gr.			
76	Rhubarb Comp., U. S. P. .	1 90	1 00	25
	.13 Gm. Rhubarb. . . . . 2 grs.			
	.10 Gm. Purified Aloes 1 1-2 grs.			
	.06 Gm. Myrrh . . . . . . 1 gr.			
	.005 Cc. Oil Peppermint 1-12 m.			
77	Salicin, 2 1-2 grs.	3 60	1 85	45
78	Salicin, 5 grs.	5 20	2 65	57
79	Salicylic acid, 2 1-2 grs. . .	1 40	75	25
80	Salicylic acid, 5 grs.	1 75	90	20

No.		Per 1,000.	Per 500.	Per 100.
81	Saline Chalybeate Tonic, Flint's	$1 10	$ 60	$ 17
	Sodium chloride . . . . . 3 grs.			
	Potassium chloride . . 3-20 gr.			
	Potassium sulphate . . 1-10 gr.			
	Potassium carbonate . . 1-20 gr.			
	Sodium carbonate . . 3-5 gr.			
	Magnes. carbonate . . 1-20 gr.			
	Prec. Calcium phos. . 1-2 gr.			
	Prec. Calcium carb. . . 1-20 gr.			
	Reduced Iron . . . . . 9-20 gr.			
	Vallet's mass. . . . . 1-20 gr.			
82	Salol, 2 1-2 grs.	4 75	2 40	53
83	Salol, 5 grs.	8 65	4 40	92
	Seiler's. See Antiseptic, "A" (21)	2 00	1 05	25
	Per pound jar, $1 50.			
84	Soda Mint	60	34	15
	Per pound jar, $0 65.			
85	Soda Mint and Pepsin . .	2 90	1 50	35
	Per pound jar, $4 00.			
	Sodium bicarbonate . . . . 4 grs.			
	Pepsin . . . . 1 gr.			
	Oil Peppermint . . . . . q. s.			
86	Sodium bicarbonate, 5 grs. .	50	30	15
	Per pound jar, $0 60.			
87	Sodium bicarbonate, 10 grs. .	60	34	15
88	Sodium bromide, 5 grs. .	1 40	75	20
89	Sodium bromide, 10 grs. . .	2 35	1 20	30
90	Sodium salicylate, 3 grs. . .	1 75	90	25
91	Sodium salicylate, 5 grs. . .	2 10	1 10	30
92	Strychnine, 1-120 gr. . . .	95	50	15
93	Strychnine, 1-60 gr. . . .	95	50	15
94	Sulphurated Antimony, 1-4 gr.	2 50	1 30	30
95	Terpin hydrate, 2 grs. . . .	1 90	1 00	25
96	Terpin hydrate, 3 grs. . . .	2 40	1 25	30
97	Terpin hydrate, 5 grs. . . .	3 40	1 75	40
98	Zinc phosphide Compound .	1 50	80	20
	Zinc phosphide . . . . 1-10 gr.			
	Ext. Nux Vomica . . . . 1-4 gr.			

LOZENGES AND TROCHES.

The lozenges offered in the following list are made by machine compression, from the purest obtainable materials, and are warranted free from all adulterants. The quantity of the medicinal ingredient in each lozenge is stated, and in the cases of official lozenges this statement is made in terms, both of metric and apothecaries' measure. The weight of each lozenge is 20 grains, except those marked with an asterisk (*), the weight of which will be found stated in Note 74. We will be pleased to make special quotations on lozenges made from private formula in lots of not less than five pounds.

No.		Per ℔. in glass jars.
1.	*Alum and Kino	$ 70
	Alum . . . . 1 gr.	
	Kino . . . . 1 gr.	
2.	Alum and Rose	60
	Alum . . . . 1 gr.	
	Rose leaves . . . . 1 gr.	
3.	*Ammonia, Dr. Jackson's	1 10
	Ammonium chloride . . . . . 1-2 gr.	
	Morphine hydrochlor. . . . 1-60 gr.	
4.	*Ammonium chloride, "A," U. S. P. .	60
	.10 Gm. Ammon. chloride. 1 1-2 grs.	
	.25 Gm. Ext. Licorice , 4 grs.	
5.	*Ammonium chloride, "B"	80
	Ammonium chloride . . . . 2 grs.	
	Extract Licorice . . . . . 8 grs.	
6.	Ammonium chloride, Cubeb and Licorice	90
	Ammonium chloride. . . . 2 grs.	
	Cubeb . . . . 1 gr.	
	Extract Licorice. . . . . 7 grs.	
	Benzoic acid, 2 grs. (105)	75
7.	Bismuth and Charcoal	1 00
	Bismuth subnitrate . . . . . 2 grs.	
	Charcoal . . . . 2 grs.	
8.	Bronchial, "A," Stearns	90
	Oleoresin Cubeb . . . . 1-8 gr.	
	Oil Sassafras . . . . . 1-16 gr.	
	Extract Licorice . . . . 2 1-2 grs.	

No.		Per lb. in glass jars.

9. Bronchial, "B" . . . . $ 50
    Cubeb . . . . 1-2 gr.
    Extract Licorice . . . . 1 gr.
    Balsam Tolu . . . . 1-10 gr.
    Oil Sassafras . . . . 1-20 gr.

10. Bronchial, "C" 50
    Cubeb . . . . 3-8 gr.
    Extract Licorice . . . . . . 4-5 gr.
    Oil Sassafras . . . . 1-40 gr.

11. Brown Mixture 65
    Opium . . . . 1-20 gr.
    Extract Licorice . . . . . 3 grs.
    Benzoic acid . . . . 1-20 gr.
    Camphor . . . . 1-20 gr.
    Tartar Emetic . . . . . . 1-40 gr.
    Oil Anise . . . . 1-20 gr.

12. Brown Mixture and Ammonium chloride. 65
    Brown Mixture . . . . . . 85 m.
    Ammonium chloride . . . . 3 grs.

13. Capsicum, 1-5 gr 65
14. Carbolic acid, 1-2 gr 50
15. Carbolic acid, 1 gr 50
16. *Catechu, U. S. P., .06 Gm. (1 gr.) . . 75
17. *Chalk, U. S. P. 60
    .25 Gm. Prepared Chalk. (4 grs.)
18. Charcoal, 4 grs 50
19. *Charcoal, 10 grs 75
    Without sugar.
20. *Charcoal, 20 grs 65
    Without sugar.
21. Chlorodyne 1 00
    Balsam Tolu . . . . 1-10 gr.
    Lobelia . . . . 1-3 gr.
    Cannabis Indica . . . . . 2-3 gr.
    Morphine sulphate . . . . 1-60 gr.
    Tartar emetic . . . . . . 1-60 gr.
22. Cinnamon, 20 grs 75
23. Cloves, 20 grs 75
24. Cocaine hydrochlorate, 1-6 gr . . . 3 00
25. Cocaine hydrochlorate, 1-4 gr . . . 4 00
26. Coryza, Spitta's. (Pastilles de Paris) 1 00
    Oleoresin Cubeb . . . . . 1-5 gr.
    Balsam Tolu . . . . 1-5 gr.
    Oil Sassafras . . . . . . 1-10 gr.
    Extract Licorice . . . . . 7 grs.

		Per ℔. in glass jars.

No.

27. **Cubeb, U. S. P. $1 25
    .04 Gm. Oleoresin Cubeb. . . 3-5 gr.
    .01 Cc. Oil Sassafras . . . . 1-7 gr.
    .25 Gm. Ext. Glycyrrhiza. . 3 4-5 grs.

28. Dover's Powder, 1 gr. 70
    Powder of Ipecac and Opium.

29. Dover's Powder, 2 1-2 grs. 75
30. Dover's Powder, 5 grs. 90
31. Eucalyptus, 3 grs. 75
32. *Ginger, "A," U. S. P. 60
    .2 Cc. Tinct. Ginger . . . . . . 3 m.

33. Ginger, "B," Strong 80
    Tinct. Ginger . . . . 15 m.

34. Ginger and Soda 75
    Tinct. Ginger . . . . 10 m.
    Sodium bicarbonate. . . . . . 2 grs.

35. Guaiac Resin, 2 grs. 60
36. *Ipecac, U. S. P., .02 Gm. (1-3 gr.) . . 75
    Ipecac and Opium.   See Dover's Powder!

37. Iron, U. S. P. 60
    .3 Gm. Ferric hydrate. . . . . 5 grs.

38. *Iron, Reduced, B.P., 1 gr. 60
    Jackson's Ammonia.   See Ammonia (3)! 1 10
    Jackson's Pectoral.   See Pectoral (46) 1 10
    Kermes Mineral.   See Sulphurated Anti-
      mony.

39. *Licorice Extract, 5 grs. 1 10
    Without excipient.

40. *Licorice Extract, 10 grs. 1 00
    Without excipient.

41. Licorice Comp., 20 grs. 60
    Comp. Powder of Glycyrrhiza, U. S. P.

42. *Licorice and Opium, U. S. P. . . . 1 00
    .15 Gm. Ext. Glycyrrhiza . 2 1-3 grs.
    .005 Gm. Powdered Opium . 1-13 gr.

43. *Magnesia, 3 grs. 80
44. *Morphine and Ipecac, U. S. P. . . . 80
    .0016 Gm. Morphine sulphate 1-40 gr.
    .005 Gm. Ipecac . . . . . 1-13 gr.

45. Paregoric 60
    Camph. Tinct. Opium . . . . 6 m.

	Per ℔. in glass jars.
No.	
46. Pectoral, Jackson's	$1 10
Ipecac . . . . 1-15 gr.	
Sulphurated Antimony . . . 1-15 gr.	
Morphine hydrochlor. . . . 1-20 gr.	
Extract Licorice . . . . 5 grs.	
Balsam Tolu . . . . 1-6 gr.	
Oil Sassafras . . . . 1-15 gr.	
47. Peppermint, U. S. P.	65
.01 Cc. Oil Peppermint . . . . . 1-6 m.	
48. Pepsin (1:3000), 1 gr.	90
49. Pepsin, saccharated, 5 grs.	90
50. Pepsin, saccharated, 10 grs.	1 10
51. Pepsin, saccharated, 15 grs.	1 30
52. Pepsin and Bismuth	1 40
Saccharated Pepsin . . . . . 5 grs.	
Bismuth subnitrate . . . . . . 2 grs.	
53. Pepsin, Bismuth and Ginger	1 40
Saccharated Pepsin . . . . . . 2 grs.	
Bismuth subnitrate . . . . . 3 grs.	
Ginger . . . . 1 gr.	
54. Pepsin and Calcium lactophosphate . . .	1 20
Saccharated Pepsin . . . . . 3 grs.	
Calcium lactophosphate . . . . 2 grs.	
55. Pepsin and Ginger Comp.	1 10
Saccharated Pepsin . . . . . . 2 grs.	
Charcoal . . . . 2 grs.	
Magnesia . . . . 2 grs.	
Ginger . . . . 1 gr.	
56. Pepsin and Iron	1 00
Saccharated Pepsin . . . . . . 2 grs.	
Iron pyrophosphate . . . . . . 2 grs.	
58. Potassium chlorate, 2 grs. (with lemon) .	50
59. Potassium chlorate, 2 grs. (with vanilla) .	50
60. Potassium chlorate, 2 grs. (with winter-green)	50
61. Potassium chlorate, 3 grs. (with lemon) .	50
62. Potassium chlorate, 3 grs. (with vanilla) .	50
63. *Potassium chlorate, 5 grs. (with lemon)	55
64. *Potassium chlorate, 5 grs. (with vanilla)	55

		Per ℔. in glass jars.
No.		

65. 1 Potassium chlorate and Ammonium chlor-1
ide . . . . $ 60
    Potassium chlorate . . . . . 1-2 grs.
    Ammonium chloride. . . . 1-2 grs.

66. *Quinine tannate, "A," per 100, $ 1.60.
    Quinine tannate. . . . . 1 gr.
    Chocolate. . . . . 9 grs.

67. *Quinine tannate, "B," per 100, $1.20.
    Quinine tannate . . . . . . 1-2 grs.
    Chocolate . . . . 7 1-2 grs.

68. *Rhatany, Ext., U. S. P., .06 Gm. (1 gr.)   80

69. Rhatany, Cubeb and Potassium chlorate.   80.
    1 Extract Rhatany . . . . . 1 gr.
    Cubeb . . . . 1-4 gr.
    Potassium chlorate . . . . 2 grs.

70. 1 Rhubarb, 46 grs. 1 25
    "Rhubarb Blocks."
    In 1 ℔. glass-top boxes. See note 93 a.

71. Rhubarb and Ginger 80
    Rhubarb . . . . 2 grs.
    Ginger . . . . 1 gr.

72. Rhubarb, Ginger and Soda 80
    Rhubarb . . . . 2 grs.
    Ginger . . . . 2 grs.
    Sodium bicarbonate . . . . 1 gr.

73. Rhubarb and Magnesia. 80
    Rhubarb . . . . 2 grs.
    Magnesia. . . . . 2 grs.

74. Rose. 75
    Rose and Alum.   See Alum and Rose (2)   60

75. Salicylic acid, 3 grs. 1 00

76. Santonin, 1-4 gr. 50

77. Santonin, 1-4 gr. (with chocolate) . . 60

78. Santonin, 1-2 gr. 70

79. Santonin, 1-2 gr. (with chocolate) . . 80

80. *Santonin, 1 gr. (pink) 90

81. *Santonin, 1 gr. (with chocolate) . . 1 00

82. Santonin and Calomel, "A" 55
    Santonin . . . . 1-4 gr.
    Mild Mercurous chloride . 1-4 gr.

83. Santonin and Calomel, "B" 65
    Santonin . . . . 1-4 gr.
    Mild Mercurous chloride . 1-4 gr.
    Chocolate . . . . q. s.

No.		Per ℔. in glass jars.
84.	Santonin and Calomel, "C" . . . . $	75
	Santonin . . . . 1-2 gr.	
	Mild Mercurous chloride . . . . 1-2 gr.	
85.	Santonin and Calomel, "D".	80
	Santonin . . . . 1-2 gr.	
	Mild Mercurous chloride . . . 1-2 gr.	
	Chocolate . . . . q. s.	
86.	Santonin, Calomel and Podophyllin, "A"	80
	Santonin . . . . 1-2 gr.	
	Mild Mercurous chloride . . . 1-2 gr.	
	Resin Podophyllum . . . . . 1-20 gr.	
87.	Santonin, Calomel and Podophyllin, "B"	85
	Santonin . . . . 1-2 gr.	
	Mild Mercurous chloride . . 1-2 gr.	
	Resin Podophyllum . . . . . 1-20 gr.	
	Chocolate . . . . q. s.	
88.	Santonin and Podophyllin, "A" .	75
	Santonin . . . . 1-2 gr.	
	Resin Podophyllum . . . . . 1-20 gr.	
89.	Santonin and Podophyllin, "B"	80
	Santonin . . . . 1-2 gr.	
	Resin Podophyllum . . . . 1-20 gr.	
	Chocolate . . . . q. s.	
90.	*Sodium bicarb., U. S. P., 2 Gm. (3 grs.)	50
91.	*Sodium santoninate, 1 gr.	1 00
92.	Sulphur, 7 1-2 grs.	50
93.	Sulphur and Cream of Tartar	60
	Sulphur . . . . 5 grs.	
	Potassium bitartrate . . . . 1 gr.	
94.	Sulphurated Antimony, 1-4 gr.	75
95.	Sulphurated Antimony, 1-2 gr. . . .	75
96.	*Tannic acid, B. P., 1-2 gr.	70
97.	*Tannic acid, U. S. P., .06 Gm. (1 gr.) .	75
98.	Tar Compound, "Pine Tree"	60
	Extract Licorice . . . . 1-3 gr.	
	Potassium chlorate . . . . . 2-3 gr.	
	Cubeb . . . . 1-3 gr.	
	Tar . . . . 1-8 gr.	
99.	Tar and Wild Cherry	60
	Tar . . . . 1-40 gr.	
	Wild Cherry . . . . 2 grs.	
	Squill . . . . 3-20 gr.	
	Lobelia . . . . 1-10 gr.	
	Ipe a . . . . 1-10 gr.	
	Tincture Opium . . . . 4-5 m.	

No.		Per ℔. in glass jars.
100.	Tolu, Tincture, 10 m. . . . . $	60
101.	White Pine Compound 1	00
	Wild Cherry . . . . 3 grs.	
	White Pine . . . . 3 grs.	
	Balm of Gilead buds . . . . 3-8 gr.	
	Spikenard. . . . . 3-8 gr.	
	Sassafras. . . . . 3 3-16 grs.	
	Morphine acetate . . . . . 1-100 gr.	
102.	Wild Cherry Compound	80
	Wild Cherry . . . . 1-10 gr.	
	Ipecac . . . . 1-50 gr.	
	Morphine sulphate . . . . . . 1-50 gr.	
	Sulphurated Antimony . . . . 1-4 gr.	
	Oil Bitter Almonds . . . . . 1-100 gr.	
	Tincture Veratrum Viride . . 3-10 gr.	
103.	Wintergreen	60
104.	Wistar's Cough	90
	Powdered Opium . . . . . 1-10 gr.	
	Extract Licorice . . . . 2 grs.	
	Oil Anise . . . . 1-30 gr.	
105.	Benzoic acid, 2 grs.	75

ALKALOIDS.

Although we do not manufacture a large variety of alkaloids we are in position to supply very fine grades of those enumerated below. At the same time we are able to offer very close figures to large consumers and we solicit correspondence for special quotations on quantities.

	Per oz.
Berberine hydrochlorate $3	00
Yellow alkaloid of Golden Seal.	
Berberine phosphate 3	25
Yellow alkaloid of Golden Seal.	
Berberine sulphate 2	00
Yellow alkaloid of Golden Seal.	
Hydrastine 7	00
White alkaloid of Golden Seal.	
Hydrastine hydrochlorate 7	00
White alkaloid of Golden Seal.	
Sanguinarine nitrate, per vial of 15 grs., $1.50.	
Sanguinarine sulphate, per vial of 15 grs., $1.50.	
Veratrine 4	00

FILLED CAPSULES.

Capsules afford a very convenient and satisfactory means of administering disagreeable medicines. The ready solubility of the gelatin in the fluids of the stomach liberates the contents of the capsule very quickly and permits the assimilative processes to begin without delay, thereby conducing to prompt action of the medicine.

Our filled capsules are strictly true to formula and are not offered in competition with those which are prepared from inferior materials, and are of short measure.

Unless otherwise directed, we shall send all filled capsules in boxes of 12, except Extra Sized Soft Elastic, which will be sent in boxes of 6 capsules.

HARD-FILLED CAPSULES.

No.		Per dozen boxes 12 in box.	Per dozen boxes 24 in box.	Per 100 in bulk.
1	Apiol . . . . Apiol. . . . . 4 m. Olive Oil. . . . . . . 6 m. See also Soft Elastic Capsules.	$3 30	$6 30	$2 00
2	Cascara Sag. Ext., 2 grs. .	1 35	2 40	70
3	Cascara Sag. Ext., 3 grs. .	1 50	2 70	85
5	Castor Oil, 10 m. . . . . . See also Soft Elastic and Extra- sized Capsules.	1 20	2 10	60
6	Castor Oil and Podophyllin. Castor Oil. . . . . . . 10 m. Resin Podophyllum . . 1-8 gr. See also Soft Elastic Capsules.	1 20	2 10	60
7	Cod Liver Oil, Norweg. 10 m. See also Soft Elastic and Extra- sized Capsules.	1 20	2 10	60
8	Cod Liver Oil and Iron . . Cod Liver Oil, Norwegian 10 m. Ferrous iodide . . . . . 1-2 gr. See also Soft Elastic Capsules.	1 35	2 40	70

No.		Per dozen boxes 12 in box.	Per dozen boxes 24 in box.	Per 100 in bulk.
9	Copaiba, Para, 10 m. . . . . See also Soft Elastic Capsules.	$1 20	$2 10	$ 60
10	Copaiba and Cubeb Oleoresin Copaiba, Para . . . . . J 8 m. Oleoresin Cubeb . . . . . . 2 m. See also Soft Elastic Capsules.	1 80	3 30	1 00
11	Copaiba and Cubeb Oil, black Copaiba, Para . . . . . . 7 m. Oil Cubeb . . . . 3 m. See also Soft Elastic Capsules.	1 80	3 30	1 00
12	Copaiba, Cubeb and Sandal . Copaiba, Para. . . . . . . 6 m. Oil Cubeb . . . . 2 m. Oil Sandal, East India . . 2 m. See also Soft Elastic Capsules.	2 30	4 30	1 35
13	Copaiba and Iron . . . . . Iron and Ammon. citrate. 2 grs. Copaiba, Para . . . . . . 9 m.	1 35	2 40	70
14	Copaiba and Sandal . . . . Copaiba, Para . . . . . . 7 m. Oil Sandal, East India . . 3 m. See also Soft Elastic Capsules.	2 25	4 20	1 30
15	Eucalyptus Oil . . . . . . Oil Eucalyptus . . . . . . 5 m. Olive Oil . . . . 5 m.	1 35	2 40	70
16	Haarlem Oil, 10 m. . . . , . . .	1 20	2 10	60
19	Male Fern and Kamala . . Oleoresin Aspidium. . . . 7 m. Kamala, sifted . . . . . . 4 grs. See also Soft Elastic Capsules.	3 50	6 70	2 10
22	Sandal Oil, East India, 10 m. See also Soft Elastic Capsules.	3 30	6 30	2 00
23	Sandal and Cassia . . . . . . Oil Sandal, East India , 9 m. Oil Cassia . . . . 1 m. See also Soft Elastic Capsules.	3 00	5 70	1 80
26	Tar, purified, 10 m. . . . . . See also Soft Elastic Capsules.	1 35	2 45	70
28	Turpentine Oil, 10 m. . . . See also Soft Elastic Capsules.	1 35	2 45	70

SOFT ELASTIC CAPSULES.

No.		Per dozen boxes 12 in box.	Per dozen boxes 24 in box.	Per 100 in bulk.
29	Apiol . . . . Apiol. . . . . 5 m. Olive Oil . . . . 5 m. See also Hard Filled Capsules.	$3 60	$6 90	$2 20
	Balsam Fir, 10 m, (58) . .	1 35	2 40	70
	Balsam Peru, 10 m. (59). .	1 65	3 00	90
31	Salol Compound	4 00	7 70	2 45
32	Castor Oil, 10 m. . . . . . . See also Hard Filled and Extra Sized Capsules.	1 20	2 10	60
33	Castor Oil and Podophyllin . Castor Oil. . . . . 10 m. Resin Podophylium. . . 1-8 gr. See also Hard Filled Capsules.	1 20	2 10	60
34	Chloroform, 10 m.	2 40	4 50	1 40
35	Cod Liver Oil, Norwegian, 10 m. . . . . See also Hard Filled and Extra Sized Capsules.	1 20	2 10	60
36	Cod Liver Oil and Iodine . Cod Liver Oil, Norwegian 10 m. Iodine . . . . 1-4 gr.	1 35	2 40	70
37	Cod Liver Oil and Iron . . . Cod Liver Oil, Norwegian 10 m. Ferrous iodide . . . . . 1-2 gr. See also Hard Filled Capsules.	1 35	2 40	70
38	Cod Liver Oil and Phosphorus Cod Liver Oil, Norwegian 10 m. Phosphorus . . . . 1-60 gr.	1 20	2 10	60
39	Copaiba, Para, 10 m. . . . . See also Hard Filled Capsules.	1 20	2 10	60
40	Copaiba and Cubeb Oleoresin Copaiba, Para . . . . . . 7 m. Oleoresin Cubeb . . . . . 3 m. See also Hard Filled Capsules.	1 80	3 30	1 00

CAPSULES.

No.	·ÉTIC CAPSU	Per dozen boxes 12 in box.	Per dozen boxes 24 in box.	Per 100 in bulk.
41	Copaiba and Cubeb Oil. . . Copaiba, Para . . . . . . 7 m. Oil Cubeb . . . . 3 m.	$2 00	$3 70	$1 15
42	Copaiba and Cubeb Oil. . . Oil Copaiba . . . . . . 6 m. Oil Cubeb . . . . 4 m.	2 00	3 70	1 15
43	Copaiba, Cubeb and Sandal . Copaiba, Para . . . . '. . 6 m. Oil Cubeb . . . . 2 m. Oil Sandal, East India . . 2 m.	2 40	4 50	1 40
44	Copaiba, Cubeb and Turpen- tine . . . . Copaiba, Para . . . . . . 4 m. Oil Cubeb . . . . 2 m. Oil Turpentine . . . . . . 4 m.	1 80	3 30	1 00
45	Copaiba and Sandal . . . . Copaiba, Para . . . . . . 5 m. Oil Sandal, East India . . 5 m.	2 40	4 50	1 40
46	Cubeb Oil, 10 m.	3 75	7 20	2 30
47	Cubeb Oleoresin, 10 m. . .	3 00	5 70	1 80
48	Cubeb and Sandal . . . . . Oil Cubeb . . . . 5 m. Oil Sandal, East India . 5 m.	3 25	6 20	1 95
49	Male Fern and Kamala . . Oleoresin Aspidium . . . 7 m. Kamala, sifted . . . . . . 4 grs.	4 00	7 70	2 45
	Salol Compound (31) . . . Salol. . . . . 3 1-2 grs. Oleoresin Cubeb . . . . . 5 m. Copaiba . . . . 10 m. Pepsin . . . . 1 gr.	4 00	7 70	2 45
50	Sandal Oil, East India, 10 m.	3 60	6 90	2 20
51	Sandal and Cassia . . . . Oil Sandal, East India . . 9 m. Oil Cassia . . . . 1 m.	3 25	6 20	1 95
52	Tar, purified, 10 m.	1 20	2 10	60
53	Terebene, 10 m.	2 00	3 70	1 15
54	Turpentine Oil, 5 m. . . .	1 20	2 10	60
55	Wintergreen Oil, 10 m. . .	2 40	4 50	1 40

SOFT ELASTIC CAPSULES.

No.		Per dozen boxes 12 in box.	Per dozen boxes 24 in box.	Per 100 in bulk.
56	Wintergreen Oil, 5 m. . . .	$1 80	$3 30	$1 00
57	Wormseed Oil	1 80	3 30	1 00
	Oil Wormseed . . . . . 2 m.			
	Olive Oil . . . . 8 m.			
58	Balsam Fir, 10 m.	1 35	2 40	70
59	Balsam Peru, 10 m.	1 65	3 00	90

EXTRA SIZED SOFT ELASTIC CAPSULES.

No.		Per dozen boxes 6 in box.	Per dozen boxes 12 in box.
60	Castor Oil, 2.5 Gm.	$1 50	$2 70
61	Castor Oil, 5 Gm.	1 80	3 30
62	Castor Oil, 10 Gm.	2 40	4 50
63	Castor Oil, 15 Gm.	3 00	5 70
64	Cod Liver Oil, Norwegian, 2.5 Gm.	1 50	2 70
65	Cod Liver Oil, Norwegian, 5 Gm. .	1 80	3 30
66	Cod Liver Oil, Norwegian, 10 Gm.	2 40	4 50
67	Cod Liver Oil, Norwegian, 15 Gm.	3 00	5 70
70	Male Fern and Castor Oil	3 00	5 70
	Oleoresin Aspidium. . . 1.0 Gm.		
	Castor Oil. . . . . . 1.5 Gm.		
71	Male Fern, Kamala and Castor Oil.	3 00	5 70
	Oleoresin Aspidium. . . 1.0 Gm.		
	Kamala . . . . 0.5 Gm.		
	Castor Oil. . . . . . 1.0 Gm.		
72	Santonin and Castor Oil	2 10	3 90
	Santonin . . . . 0.25 Gm.		
	Castor Oil. . . . . . 2.25 Gm.		

SOFT ELASTIC CAPSULES FILLED WITH POWDERS.

No.		Per dozen boxes 12 in box.	Per dozen boxes 24 in box.	Per 100 in bulk.
100	Quinine sulphate, 1 gr.			
101	Quinine sulphate, 2 grs.	Prices on application.		
102	Quinine sulphate, 3 grs.			
103	Quinine sulphate, 4 grs.			
104	Quinine sulphate, 5 grs.			
105	Quinine, Iron and Nux Vomica. . . . . Quinine sulphate. . . . 1 gr. Ferrous sulphate . . . . 1 gr. Ext. Nux Vomica . . .1-8 gr.	$1 65	$3 00	$1 00
106	Quinine and Capsicum. . . Quinine sulphate . . . 3 grs. Capsicum . . . . . .1-3 gr.	1 75	3 10	1 10
107	Acetanilid, 3 grs.	1 20	2 10	60
108	Acetanilid, 5 grs.	1 35	2 35	75
109	Antifebrin, 3 grs.	2 35	4 50	1 20
110	Antifebrin, 5 grs.	3 25	6 50	1 85
111	Antipyrine, 3 grs.	7 00	12 50	4 50
112	Antipyrine, 5 grs.	10 00	18 50	6 50
113	Salol, 2 1-2 grs.	2 10	4 00	1 25
114	Salol, 5 grs.	2 70	5 00	1 50
115	Sulphonal, 5 grs.	15 00	25 00	8 00

ACIDS.
No charge for containers.

Hydriodic acid, 10.%, per oz. . . . . $ 35
   Ten times as strong as the U. S. P. syrup.
   See note 60.

Hydrobromic acid, diluted, 10.%, per lb. 1 50
Phosphoric acid, diluted, 10.%, per lb. 40
Phosphoric acid, 50.%, per lb. 1 60

MEDICINAL ELIXIRS.

Although elixirs have received but scant recognition in the Pharmacopœias, they are, when properly prepared, valuable additions to the list of pharmaceutical preparations. While we have ever taken pride in the elegance and palatability of our elixirs, we have not suffered their medicinal value to be made secondary to less important requirements. With these as with all other remedial agents, therapeutic worth is the great desideratum and should not be sacrificed to palatability or appearance. We feel, therefore, that we may justly claim that our elixirs are unsurpassed for therapeutic effectiveness, and that no others which are true to formula will be found more pleasing in appearance and taste.

They are put up in full measure 16 ounce and half-gallon bottles, no extra charge being made for containers. Unless otherwise sp   edicinal ingredients are stated in grains per fl   e.

No.		Per doz. pints.	Per gallon.
1.	Adjuvant, N. F.	$ 6 50	$3 35
2.	*Discontinued.*		
3.	Aloin, Strychnine and Belladonna, sweet.	8 00	4 25
	Aloin . . . . . 1 3-5 grs.		
	Ext. Belladonna leaves . . 1 gr.		
	Strychnine . . . . . 8-60 gr.		
	Tonic laxative.		
4.	Ammonium bromide, 40 grs.	9 00	5 00
	Sedative.		
5.	Ammonium valerianate, 16 grs.	9 00	5 00
	Nervine.		
6.	Ammonium and Morphine valerianates.	12 00	7 00
	Ammonium valerianate . . 16 grs.		
	Morphine valerianate. . . 1 gr.		
	Nerve sedative.		

MEDICINAL ELIXIRS.

No.		Per doz. pints.	Per gallon
7.	Ammonium and Quinine valerianates	$11 25	$6 50
	Ammonium valerianate . . 16 grs.		
	Quinine valerianate. . . . . 4 grs.		
	Nerve tonic.		
8.	Aromatic, colorless	6 00	3 00
9.	Aromatic, red	6 00	3 00
10.	Asthmatic	12 00	7 50
	Quebracho . . . . 45 grs.		
	Coffee . . . . 45 grs.		
	Ipecac. . . . . 8 grs.		
	Blood-root . . . . 4 grs.		
	Potassium iodide . . . . . 16 grs.		
	Potassium nitrite . . . . 16 grs.		
	Chloroform . . . . q. s.		
11.	Bark and Iron	7 50	4 00
	Peruvian bark . . . . . . 5 grs.		
	Iron protoxide . . . . . . 6 grs.		
	Tonic, hematic.		
12.	Berberine and Iron	12 00	7 00
	Berberine sulphate . . . . . 2 grs.		
	Ferric pyrophosphate. . . . 8 grs.		
	Ferruginous tonic.		
13.	Bismuth and Ammon. cit. 16 grs.	9 00	5 00
	Stomachic, astringent.		
	Bismuth and Pepsin. See Pepsin and Bismuth.		
14.	Bismuth and Strychnine . . .	9 00	5 00
	Bismuth and Ammon. citrate 16 grs.		
	Strychnine. . . . . 8-60 gr.		
	Stomachic, tonic.		
	Black Cohosh Comp. See Cimicifuga Comp. (51)	8 00	4 25
15.	Buchu, 60 grs.	9 00	5 00
	Diuretic.		
16.	Buchu Comp.	10 50	6 00
	Buchu. . . . . 30 grs.		
	Cubeb . . . . 30 grs.		
	Juniper . . . . 30 grs.		
	Uva Ursi . . . . 30 grs.		
	Spirit Nitrous Ether . . . 30 m.		
	Diuretic, antiblennorrhagic.		

No.		Per doz. pints.	Per gallon.
17.	Buchu, Juniper and Potassium acetate	$9 75	$5 50
	Buchu . . . . 45 grs.		
	Juniper . . . . 12 grs.		
	Potassium acetate . . . . 16 grs.		
	Diuretic.		
18.	Buchu and Pareira	9 00	5 00
	Buchu . . . . 60 grs.		
	Pareira . . . . 60 grs.		
	Diuretic.		
19.	Buchu and Pareira Comp. . . .	9 00	5 00
	Buchu . . . . 30 grs.		
	Dandelion . . . . 30 grs.		
	Juniper berries . . . . . 20 grs.		
	Stone-root . . . . 20 grs.		
	Pareira . . . . 20 grs.		
	Potassium acetate . . . . 16 grs.		
	Diuretic, antilithic.		
20.	Buckthorn bark, 120 grs. . . .	9 00	5 00
	Laxative.		
21.	Buckthorn and Rhubarb . . .	10 50	6 00
	Buckthorn . . . . 60 grs.		
	Rhubarb . . . . 30 grs.		
	Laxative, mild astringent.		
22.	Caffeine, citrated, 8 grs. . . .	11 25	6 50
	Cardiac stimulant, diuretic.		
	See note 20.		
23.	Caffeine and Potass. bromide, .	11 25	6 50
	Caffeine . . . . 4 grs.		
	Potassium bromide . . . . 80 grs.		
	Cardiac stimulant, sedative.		
24.	Calisaya and Bismuth	9 00	5 00
	Calisaya bark (alkaloids) . 40 grs.		
	Bismuth and Ammon. citrate. 8 grs.		
	Stomachic, tonic.		
25.	Calisaya and Iron, "A" . . .	7 50	4 00
	Calisaya bark (alkaloids) . 40 grs.		
	Ferricyphosphate . . . . . 8 grs.		
	Tonic, hematic.		
26.	Calisaya and Iron, "B" . . .	7 50	4 00
	Calisaya bark (alkaloids), 40 grs.		
	Iron protoxide . . . . . . 8 grs.		
	Tonic, hematic.		
27.	Calisaya and Iron, "C" . . .	7 50	4 00
	Calisaya bark (alkaloids) . 40 grs.		
	Ferric pyrophosphate . . . . 8 grs.		
	Tonic, hematic.		

No.		Per doz. pints.	Per gallon.
28.	Calisaya, Iron and Bismuth . .	$ 9 00	$ 5 00
	Calisaya bark (alkaloids) . 40 grs.		
	Ferric pyrophosphate . . . 16 grs.		
	Bismuth and Ammon. citrate 8 grs.		
	Hematic tonic, mild astringent. . .		
29.	Calisaya, Iron, Bismuth and Pepsin	15 00	9 00
	Calisaya bark (alkaloids) . 40 grs.		
	Ferric pyrophosphate . . . 16 grs.		
	Bismuth and Ammon. citrate 8 grs.		
	Saccharated Pepsin . . . . 40 grs.		
	Tonic, digestant.		
30.	Calisaya, Iron, Bismuth, Pepsin and Strychnine	15 00	9 00
	Calisaya bark (alkaloids) . 40 grs.		
	Ferric pyrophosphate . . . 16 grs.		
	Bismuth and Ammon. citrate 8 grs.		
	Saccharated Pepsin . . . . 40 grs.		
	Strychnine. . . . . . . 8-60 gr.		
	Nervine, hematic tonic.		
31.	Calisaya, Iron, Bismuth and Strychnine, "A"	9 00	5 00
	Calisaya bark (alkaloids) . 40 grs.		
	Ferric pyrophosphate . . . 16 grs.		
	Bismuth and Ammon. citrate 8 grs.		
	Strychnine . . . . 8-60 gr.		
	Nervine, hematic tonic.		
32.	Calisaya, Iron, Bismuth and Strychnine, "B"	9 00	5 00
	Calisaya bark . (alkaloids) . 40 grs.		
	Iron and Ammon. citrate . . 8 grs.		
	Bismuth and Ammon. citrate 16 grs		
	Strychnine . . . . 8-60 gr.		
	Nervine, hematic tonic.		
33.	Calisaya, Iron and Pepsin . .	12 00	7 00
	Calisaya bark (alkaloids) . 40 grs.		
	Ferric pyrophosphate . . . 16 grs.		
	Saccharated Pepsin . . . . 40 grs.		
	Digestant, hematic tonic.		
34.	Calisaya, Iron and Phosphorus.	9 00	5 00
	Calisaya bark (alkaloids) . 40 grs.		
	Ferric pyrophosphate . . . 16 grs.		
	Phosphorus . . . . . . 8-100 gr.		
	Nerve tonic.		

No.	Per doz. pints.	Per gallon.

35. Calisaya, Iron and Quinine . . $10 50 $6 00
    Calisaya bark (alkaloids) . 40 grs.
    Ferric pyrophosphate . . . . 8 grs.
    Quinine sulphate . . . . . . 2 grs.
    Tonic.

36. Calisaya, Iron and Strychnine. 8 25 4 50
    Calisaya bark (alkaloids) . 40 grs.
    Ferric pyrophosphate . . . 16 grs.
    Strychnine . . . . . . 8-60 gr.
    Tonic.

37. Calisaya, Iron, Strychnine and
    Pepsin 13 50 8 00
    Calisaya bark (alkaloids) . 40 grs.
    Ferric pyrophosphate . . . 16 grs.
    Strychnine . . . . . . 8-60 gr.
    Saccharated Pepsin . . . . 40 grs.
    Tonic, digestant.

38. Calisaya, Pepsin and Bismuth. 14 25 8 00
    Calisaya bark (alkaloids) . 40 grs.
    Saccharated Pepsin . . . . 40 grs.
    Bismuth and Ammon. citrate. 8 grs.
    Tonic, digestant.

39. Calisaya, Pepsin, Bismuth and
    Strychnine 12 75 7 50
    Calisaya bark (alkaloids) . 40 grs.
    Saccharated Pepsin . . . . 40 grs.
    Bismuth and Ammon. citrate. 8 grs.
    Strychnine. . . . . . . 8-60 gr.
    Tonic, digestant.

40. Calisaya, Pepsin and Strych-
    nine 12 75 7 50
    Calisaya bark (alkaloids) . 40 grs.
    Saccharated Pepsin . . . . 40 grs.
    Strychnine . . . . . . 8-60 gr.
    Tonic, digestant.

41. Calisaya and Phosphates . . . 7 00 4 00
    Calisaya bark (alkaloids) . . 5 grs.
    Iron phosphate . . . .
    Calcium phosphate . .   of each,
    Potassium phosphate . sufficient
    Sodium phosphate . . quantities.
    Phosphoric acid . . .
    Nervine, nutrient to osseous tissue.

42. Calisaya and Strychnine . . . 9 00 5 00
    Calisaya bark (alkaloids) . 40 grs.
    Strychnine. . . . . . . 8-60 gr.
    Tonic, nerve stimulant.

No.		Per doz. pints.	Per gallon.
43.	Cascara Sagrada, 60 grs. . . . . Tonic laxative.	$ 9 00	$5 00
44.	Cascara Sagrada Comp. (Cascara Cordial).	9 00	5 00
	Cascara Sagrada . . . . . . 60 grs. Berberis Aquifolium . . . . . 30 grs. Tonic laxative.		
45.	Castanea and Belladonna . .	9 75	5 50
	Chestnut leaves . . . . . . 60 grs. Belladonna leaves . . . . . . 2 grs. Tonic, astringent, expectorant.		
46.	Cathartic Comp.	9 00	5 00
	Rhubarb . . . . 7 1-2 grs. Euonymus . . . . . . 7 1-2 grs. Mandrake . . . . 7 1-2 grs. Buckthorn . . . . 30 grs. Senna. . . . . 45 grs. Henbane. . . . . . . 2 grs. Rochelle Salt . . . . . . 30 grs. Sodium bicarbonate . . . 7 1-2 grs.		
47.	Celery Nervine	9 00	5 00
	Celery seed . . . . 32 grs. Black Haw . . . . . . 32 grs. Coca . . . . 32 grs. Nerve tonic.		
48.	Celery and Guarana	12 75	7 50
	Celery seed . . . . 60 grs. Guarana . . . . 90 grs. Nervine.		
49.	Celery and Guarana, detannated	12 75	7 50
	Celery seed . . . . 60 grs. Guarana . . . . 90 grs. Nervine.		
50.	Chloral Comp. (Chloral Anodyne).	13 50	8 00
	Chloral hydrate . . . . . 40 grs. Potassium bromide . . . 80 grs. Nux Vomica. . . . . . 8 grs. Sedative, antispasmodic.		

No.		Per doz. pints.	Per gallon.
51.	Cimicifuga Comp. . . . . $ 8 00		$4 25
	Cimicifuga . . . . 30 grs.		
	Wild Cherry . . . . . . 15 g's.		
	Senega . . . . 7 1-2 grs.		
	Blood-root . . . . . . 3 3-4 grs.		
	Ipecac . . . . 3 3-4 grs.		
	Licorice. . . . . 30 grs.		
	(Stimulating expectorant.		
52.	Cinchona (alkaloids), 40 grs. . . 10 50		6 00
	Tonic, antiperiodic.		
53.	Cinchona, ferrated 9 75		5 50
	Red Cinchona. . . . . . 40 grs.		
	Iron and Ammon. citrate . 16 grs.		
	Ferruginous tonic.		
54.	Cinchonidine, Iron and Strych-		
	nine 8 25		4 50
	Cinchonidine sulphate . . . 8 grs.		
	Ferric phosphate . . . 16 grs.		
	Strychnine . . . . . 18-60 grs.		
	Ferruginous tonic.		
55.	Coca, 120 grs. 9 75		5 50
	Nerve tonic.		
56.	Codeine sulphate, 1 gr. . . . 12 00		7 00
	Analgesic.		
57.	Corydalis Comp. 8 25		4 50
	Corydalis . . . . . . 30 grs.		
	Stillingia . . . . . . 30 grs.		
	Prickly ash . . . . . 15 grs.		
	Blue Flag . . . . . . 45 grs.		
	Potassium iodide . . . . 8 grs.		
	(Alterative.		
58.	Damiana Comp. 12 00		7 00
	Damiana . . . . 60 grs.		
	Coca . . . . 60 grs.		
	Nux Vomica . . . . . . 8 grs.		
	Phosphorus . . . . . 8-100 gr.		
	Aphrodisiac, nerve		
59.	Damiana, Nux Vomica, Iron		
	and Phosphorus 9 75		5 50
	Damiana . . . . 60 grs.		
	Phosphorus . . . . . 8-200 gr.		
	Nux Vomica . . . . . 7 1-2 grs.		
	Ferric pyrophosphate . . . 8 grs.		
	Aphrodisiac, nerve tonic.		

No.		Per doz. pints.	Per gallon.
60.	Eucalyptus Comp.	$ 9 00	$5 90
	Eucalyptus . . . . 30 grs.		
	Prickly ash . . . . 30 grs.		
	Grindelia . . . . 15 grs.		
	Gentian. . . . . 3 3-4 grs.		
	Licorice and aromatics. . . 15 grs.		
	Tonic.		
61.	Ferrated Cordial	8 00	4 25
	Ferric pyrophosphate . . . . 16 grs.		
	Tonic, stimulant.		
62.	Gentian, 20 grs.	7 50	4 00
	Tonic, stomachic.		
63.	Gentian Comp.	7 50	4 00
	Gentian. . . . . 16 grs.		
	Bitter Orange peel . . . . . 8 grs.		
	Cardamom . . . . 4 grs.		
	Tonic, stomachic.		
64.	Gentian Comp., ferrated . . .	8 00	4 25
	Gentian. . . . . 16 grs.		
	Bitter Orange peel . . . . . 8 grs.		
	Cardamom . . . . 4 grs.		
	Ferric pyrophosphate . . . . 8 grs.		
	Tonic, stomachic.		
65.	Gentian and Iron Chloride . .	7 50	4 00
	Gentian . . . . 16 grs.		
	Tincture Iron citro-chloride 16 m.		
	Ferruginous tonic.		
66.	Grindelia Aromatic	8 25	4 50
	Grindelia . . . . 60 grs.		
	Antasthmatic.		
67.	Guarana, 120 grs.	15 00	9 00
	Nerve stimulant.		
68.	Helonias Compound.	9 00	5 00
	Helonias . . . . 120 grs.		
	Buchu . . . . 30 grs.		
	Trillium. . . . . 30 grs.		
	Gentian . . . . 30 grs.		
	Hydrastis . . . . 30 grs.		
	Diuretic, genito-urinary tonic.		
	See note 58.		
69.	Iron phosphate, 16 grs. . . .	8 25	4 50
	Chalybeate.		
70.	Iron and Quinine, "A" . . .	10 50	6 00
	Ferric phosphate . . . . . 8 grs.		
	Quinine sulphate . . . . . 4 grs.		
	Tonic.		

.ELIXIRS.

MEDICINAL ELIXIRS. 165

No.	Per doz. pints.	Per gallon.
71. Iron and Quinine, "B"	$10 50	$6 00
Ferric pyrophosphate . . . 16 grs.		
Quinine hydrochlorate . . . 8 grs.		
Tonic.		
72. Iron and Quinine, "C"	10 50	6 00
Ferric phosphate . . . . . 8 grs.		
Quinine phosphate . . . . 4 grs.		
Tonic.		.16
73. Iron, Quinine and Arsenic	10 50	6 00
Ferric phosphate . . . . . 8 grs.		
Quinine sulphate . . . . 4 grs.		
Arsenous acid . . . . . . 8-16 gr.		
Tonic, antiperiodic.		
74. Iron, Quinine and Strychnine, "A"	10 50	6 00
Ferric phosphate . . . . . 16 grs.		
Quinine sulphate . . . . . 8 grs.		
Strychnine . . . . 8-60 gr.		
Nerve tonic, antiperiodic.		
75. Iron, Quinine and Strychnine, "B"	10 50	6 00
Ferric pyrophosphate . . . 16 grs.		
Quinine hydrochlorate . . . 8 grs.		
Strychnine . . . . . . 8-60 gr.		
Nerve tonic, antiperiodic.		
76. Iron, Quinine and Strychnine, "C"	10 50	6 00
Ferric phosphate . . . . . 16 grs.		
Quinine phosphate . . . . . 8 grs.		
Strychnine phosphate . . 8-60 gr.		
Nerve tonic, antiperiodic.		
77. Iron, Quinine, Strychnine and Pepsin	12 00	7 00
Ferric phosphate . . . . . 8 grs.		
Quinine sulphate . . . . . 4 grs.		
Strychnine . . . . . . 8-60 gr.		
Saccharated Pepsin . . . . 40 grs.		
Nerve tonic, digestant.		
78. Iron and Strychnine, "A"	8 25	4 50
Ferric phosphate . . . . . 16 grs.		
Strychnine . . . . . . 8-60 gr.		
Tonic.		
79. Iron and Strychnine, "B"	8 25	4 50
Ferric pyrophosphate . . . 16 grs.		
Strychnine . . . . . 8-60 gr.		
Tonic.		

No.		Per doz. pints.	Per gallon.
80.	Kola Comp.	$10 00	$5 75
	Kola .... 48 grs.		
	Celery .... 48 grs.		
	Coca .... 64 grs.		
	Nerve stimulant, conservator of energy. See note 70.		
81.	Lactinated Pepsin, 38 grs. . .	12 00	6 00
	Digestant. See note 71 a.		
82.	Lactinated Pepsin and Bismuth	12 00	7 00
	Lactinated Pepsin . . . . . 32 grs.		
	Bismuth and Ammon. citrate, 8 grs.		
	Digestant, mild astringent.		
	See note 71 a.		
83.	Lactinated Pepsin, Bismuth, and Strychnine	12 00	7 00
	Lactinated Pepsin . . . . . 32 grs.		
	Bismuth and Ammon. citrate. 8 grs.		
	Strychnine . . . . . 1-12 gr.		
	Digestant, nerve tonic.		
	See note 71 a.		
84.	Lactinated Pepsin and Calisaya.	12 00	7 00
	Lactinated Pepsin . . . . . 32 grs.		
	Calisaya bark . . . . . . 50 grs.		
	Digestant, tonic.		
	See note 71 a.		
85.	Lactinated Pepsin, Calisaya, Iron and Bismuth	12 00	7 00
	Lactinated Pepsin . . . . . 32 grs.		
	Calisaya bark . . . . . . 40 grs.		
	Ferric phosphate . . . . . 12 grs.		
	Bismuth and Ammon. citrate. 2 grs.		
	Digestant, general tonic.		
	See note 71 a.		
86.	Lactinated Pepsin, Gentian and Chloride of Iron	12 00	7 00
	Lactinated Pepsin . . . . . 32 grs.		
	Gentian . . . . . 8 grs.		
	Iron proto-chloride . . . . 8 grs.		
	Digestant, general tonic.		
	See note 71 a.		

No.		Per doz. pints.	Per gallon.
87.	Lactinated Pepsin with Iron, Quinine and Strychnine phosphates.	$12 00	$7 00
	Lactinated Pepsin . . . . . 32 grs.		
	Ferric phosphate . . . . . 4 grs.		
	Quinine phosphate . . . . 4 grs.		
	Strychnine phosphate . . 1-16 gr.		
	Digestant, nerve tonic.		
	See note 71 a.		
88.	Licorice Aromatic,	7 00	3 65
	To disguise bitter taste of quinine.		
	See note 72.		
89.	Manaca and Salicylates . . .	18 00	11 00
	Manaca . . . . 80 grs.		
	Sodium salicylate . . . . 64 grs.		
	Potassium salicylate . . . . 32 grs.		
	Lithium salicylate . . . . 8 grs.		
	Antirheumatic, antarthritic.		
90.	Matico Compound	9 00	5 00
	Matico . . . . 88 grs.		
	Cubeb . . . . 28 grs.		
	Buchu . . . . 28 grs.		
	Stimulating diuretic.		
91.	Mitchella Compound	9 00	5 00
	Mitchella . . . . 60 grs.		
	Helonias . . . . 15 grs.		
	Cramp-bark . . . . . . 15 grs.		
	Blue Cohosh. . . . . . . 15 grs.		
	Uterine tonic, emmenagogue.		
	See note 58.		
	Orange, colorless. See Aromatic (8)	6 00	3 00
	Orange, red. See Aromatic (9)	6 00	3 00
92.	Pancreatin, saccharated, 40 grs.	10 50	7 00
	Starch digestant.		
	Partridgeberry Comp. See Mitchella Comp. (91).	9 00	5 00
93.	Pepsin, saccharated, 40 grs. .	10 50	6 00
	Saccharated Pepsin, U. S. P.		
	Digestant.		
94.	Pepsin and Bismuth	12 00	7 00
	Saccharated Pepsin . . . . 40 grs.		
	Bismuth and Ammon. citrate 8 grs.		
	Digestant.		

	Per doz. pints.	Per gallon.

No.

95. Pepsin, Bismuth and Iron . . $12 00 $7 00

   Saccharated Pepsin . . . . 40 grs.
   Bismuth and Ammon. citrate 8 grs.
   Ferric pyrophosphate. . . . 8 grs.
   Tonic, digestant.

96. Pepsin, Bismuth, Iron and
   Quinine 12 75   7 50

   Saccharated Pepsin . . . 40 grs.
   Bismuth and Ammon. citrate 8 grs.
   Ferric pyrophosphate. . . . 8 grs.
   Quinine sulphate . . . . . 4 grs.
   Tonic, digestant.

97. Pepsin, Bismuth, Iron and
   Strychnine 12 00   7 00

   Saccharated Pepsin . . . 40 grs.
   Bismuth and Ammon. citrate 8 grs.
   Ferric pyrophosphate. . . . 8 grs.
   Strychnine . , ... 8-60 gr.
   Tonic, digestant.

98. Pepsin, Bismuth and Nux Vomica 12 00   7 00

   Saccharated Pepsin . . . . 40 grs.
   Bismuth and Ammon. citrate. 8 grs.
   Nux Vomica . . . . 15 grs.
   Digestant, nerve tonic.

99. Pepsin, Bismuth and Pancreatin. 12 00   7 00

   Saccharated Pepsin . . . . 40 grs.
   Bismuth and Ammon. citrate. 8 grs.
   Saccharated Pancreatin. . 40 grs.
   Digestant, mild astringent.

100. Pepsin, Bismuth and Quinine . 12 00   7 00

   Saccharated Pepsin . . . . 40 grs.
   Bismuth and Ammon. citrate. 8 grs.
   Quinine sulphate . . . . . 2 grs.
   Digestant, tonic.

101. Pepsin, Bismuth and Strychnine 12 00   7 00

   Saccharated Pepsin . . . 40 grs.
   Bismuth and Ammon. citrate. 8 grs.
   Strychnine . . . . 8-60 gr.
   Digestant, nerve tonic.

No.		Per doz. pints.	Per gallon.
102.	Pepsin, Bismuth, Strychnine and Pancreatin	$12 75	$7 50
	Saccharated Pepsin . . . . 40 grs.		
	Bismuth and Ammon. citrate. 8 grs.		
	Strychnine . . . . . . 8-60 gr.		
	Saccharated Pancreatin . . . 40 grs.		
	Digestant, nerve tonic.		
103.	Pepsin, Iron and Strychnine . . 12 00		7 00
	Saccharated Pepsin . . . . 40 grs.		
	Ferric pyrophosphate . . . . 8 grs.		
	Strychnine . . . . . . 8-60 gr.		
	Digestant, hematic tonic.		
104.	Pepsin and Pancreatin 12 75		7 50
	Saccharated Pepsin . . . . 40 grs.		
	Saccharated Pancreatin . . 40 grs.		
	Digestant.		
105.	Pepsin and Quinine 12 00		7 00
	Saccharated Pepsin . . . . 40 grs.		
	Quinine sulphate . . . . . 2 grs.		
	Digestant, stomachic tonic.		
106.	Pepsin and Strychnine 12 00		7 00
	Saccharated Pepsin . . . . 40 grs.		
	Strychnine . . . . 8-60 gr.		
	Digestant, nerve tonic.		
107.	Pepsin and Wafer Ash 13 50		8 00
	Saccharated Pepsin . . . . 40 grs.		
	Wafer Ash . . . . 60 grs.		
	Digestant, stomachic tonic.		
108.	Phosphorus, 8-100 gr. 9 00		5 00
	Nervine, nutrient to osseous tissue.		
109.	Phosphorus and Nux Vomica . 9 00		5 00
	Phosphorus. . . . . . 8-100 gr.		
	Nux Vomica . . . . 8 grs.		
	Nervine, nutrient to osseous tissue.		
110.	Phosphorus and Strychnine . . 9 75		5 50
	Phosphorus . . . . . . 8-100 gr.		
	Strychnine . . . . 8-60 gr.		
	Nervine, nutrient to osseous tissue.		
111.	Potassium bromide, 80 grs. . 9 00		5 00
	Sedative.		

No.		Per doz. pints.	Per gallon.
112.	Rhubarb and Potassa Comp.	$ 9 00	$5 00
	Rhubarb . . . . 30 grs.		
	Hydrastis . . . . 10 grs.		
	Potassium carbonate . . . 3 1-2 grs.		
	Cinnamon . . . . 15 grs.		
	Oil of Peppermint . . . . . q. s.		
	Alkaline laxative, antidiarrheic.		
113.	Salicylic acid, 16 grs. . . .	10 50	6 00
	Antirheumatic.		
114.	Sedative Compound	11 00	6 00
	Hydrastine . . . . 1 gr.		
	Black Haw . . . . 60 grs.		
	Jamaica Dogwood . . . . 30 grs.		
	Uterine sedative, antispasmodic.		
	Simple, colorless (8)	6 00	3 00
	Simple, red (9)	6 00	3 00
115.	Sodium bromide, 80 grs. . .	9 00	5 00
	Sedative.		
116.	Sodium salicylate, 40 grs. .	7 50	4 00
	Antirheumatic.		
117.	Strychnine valerianate, 8-60 gr.	7 50	4 00
	Nervine.		
118.	Taraxacum Compound. . . .	7 50	4 00
	Vehicle.		
119.	Terpin hydrate, 8 grs. . . .	9 00	5 00
	Expectorant.		
120.	Terpin hydrate and Codeine .	12 00	7 00
	Terpin hydrate . . . . 8 grs.		
	Codeine sulphate . . . . . . 1 gr.		
	Sedative, expectorant.		
121.	Valerian, 120 grs.	9 00	5 00
	Nervine.		
122.	Wahoo, 120 grs.	9 00	5 00
	Tonic.		
123.	Wild Cherry and Iron	9 00	5 00
	Wild Cherry . . . . 60 grs.		
	Ferric pyrophosphate . . . 8 grs.		
	Ferruginous tonic.		
124.	Zinc valerianate, 8 grs. . .		5 00
	Nerve tonic.		

MEDICINAL SYRUPS.

The syrups offered in the following list are those of the U. S. Pharmacopœia, with the addition of some others that have received the approbation of medical men in various parts of the country. We call special attention to the fact that our prices are for pints of 16 fluidounces (instead of for pounds, of about 12 fluid-ounces), and for gallons in half-gallon bottles. This should be borne in mind in comparing our prices with those of other houses. Unless otherwise specified the medicinal ingredients are stated in grains per fluid-ounce.

No.		Per doz. pints.	Per gallon.
1	Blackberry, U. S. P. . . . . . . 100 Cc. contain— 1 fl. oz. contains— 25 Cc. Fl. Ext. Blackberry . 120 m. Astringent.	$ 7 20	$3 80
2	Blackberry aromatic, 120 grs. . . . Astringent.	7 20	3 80
3	Buckthorn berries, 120 grs. . . . Laxative.	7 20	3 80
4	Calcium hypophos., N. F., 16 grs. Nerve tonic.	8 25	4 50
5	Calcium iodide, N. F., 40 grs. Alterative.	12 75	7 50
6	Calcium lactophos., Stearns', 16 grs. Nutrient to osseous tissue.	13 50	8 00
7	Calcium and Sodium lactophos- phates . . . . Calcium lactophosphate . . 12 grs. Sodium lactophosphate. . . 8 grs. Nutrient to osseous tissue.	13 50	8 00
8	Calcium lactophosphate Compd, with Iron . . . . Calcium lactophosphate . . 12 grs. Sodium lactophosphate. . . 6 grs. Potassium lactophosphate . 2 grs. Iron lactophosphate . . . 1 gr. Tonic, nutrient to osseous tissue.	13 50	8 00
9	Dover's Powder, N. F., 40 grs. Sedative.	9 00	5 00

No.		Per doz. pints.	Per gallon.
10	Ginger, U. S. P.	$ 7 00	$4 50
	100 Cc. represent— 1 fl.oz. represents—		
	3 Cc. Fl. Ext. Ginger . . . . 15 m.		
	Hematic. See Hypophosphites,		
	Tonic (16)	9 00	5 00
11	Hydriodic acid, U. S. P., 1.% . .	10 75	6 75
	Alterative. See note 61.		
12	Hypophosphites, U. S. P.	9 00	5 00
	100 Cc. contain— 1 fl. oz. contains—		
	4.5 Gm. Calcium hypophos .. 20 grs.		
	1.5 Gm. Sodium hypophos. . . 7 grs.		
	1.5 Gm. Potassium hypophos. 7 grs.		
	General tonic.		
13	Hypophosphites with Iron, U. S. P.	9 00	5 00
	100 Cc. contain— 1 fl. oz. contains—		
	4.5 Gm. Calcium hypophos. . 20 grs.		
	1.5 Gm. Sodium hypophos. . . 7 grs.		
	1.5 Gm. Potassium hypophos. 7 grs.		
	1. Gm. Ferrous lactate. . 4 1-2 grs.		
	General tonic.		
14	Hypophosphites Comp. with Iron		
	and Manganese	9 00	5 00
	Calcium hypophosphite . . . 16 grs.		
	Sodium hypophosphite. . . . 8 grs.		
	Potassium hypophosphite . ! 8 grs.		
	Iron hypophosphite . . . . . 2 grs.		
	Manganese hypophosphite . . 1 gr.		
	General tonic.		
15	Hypophosphites Comp. with Manganese, Quinine and Strychnine.	9 00	5 00
	Calcium hypophosphite . . . 16 grs.		
	Sodium hypophosphite. . . . 8 grs.		
	Potassium hypophosphite . . 8 grs.		
	Iron hypophosphite . . 1 gr.		
	Manganese hypophosphite . . 1 gr.		
	Quinine hypophosphite. . . 1-2 gr.		
	Strychnine hypophosphite 8-100 gr.		
	General tonic.		
16	Hypophosphites, Tonic	9 00	5 00
	Calcium hypophosphite. . . . 1 gr.		
	Potassium hypophosphite. 1 1-2 grs.		
	Iron hypophosphite . . 1 1-2 grs.		
	Manganese hypophosphite . . 1 gr.		
	Quinine hypophosphite . . . 1-4 gr.		
	Strychnine hypophosphite. 8-100 gr.		
	See note 100.		

No.		Per doz. pints.	Per gallon.
17	Ipecac, U. S. P. . . . . 100 Cc. represent— 1 fl.oz. represents— 7 Cc. Fl. Ext. Ipecac . . . . .34 m. Expectorant.	$ 9 75	$5 50
18	Iron iodide, U. S. P. . . . . . . 10 per cent. Ferrous iodide by weight. Tonic, alterative.	12 00	7 50
19	Iron and Manganese iodides . . . Ferrous iodide . . . . . . .24 grs. Manganese iodide . . . . . .12 grs. Tonic, alterative.	13 50	8 00
20	Iron, Quinine and Strychnine phosphates, U. S. P. . . . . 100 Cc. contain— 1 fl. oz. contains— 2. Gm. Soluble Ferric phos. 9 grs. 3. Gm. Quinine sulphate . 14 grs. .02 Gm. Strychnine. . . . 1-11 gr. Tonic.	12 75	7 50
21	Iron, saccharated, N. F. . . . . . 1 fl. oz. represents 6 1-2 grs. metallic iron. Hematic tonic.	11 50	6 50
22	Lactinated Pepsin with Phosphates Lactinated Pepsin . . . . .32 grs. Iron phosphate . . . . . . .8 grs. Calcium phosphate . . . . .8 grs. Sodium phosphate . . . . . .8 grs. Potassium phosphate . . . .8 grs. Tonic, digestant. See note 71 a.	12 00	7 00
23	Lactucarium, U. S. P. . . . . . 100 Cc. represent— 1 fl.oz. represents— 10 Cc. Tinct Lactucarium . .48 m. Mild sedative.	16 50	10 00
24	Lemon.	11 00	6 00
25	Licorice Aromatic . . . . See note 72. . .	8 25	4 50
26	Manganese iodide, 16 grs. . . . . Alterative, tonic.	16 50	10 00
27	Mitchella Compound . . . . . . Mitchella. . . . . .60 grs. Helonias . . . . .15 grs. Cramp-bark . . . . . . .15 grs. Blue Cohosh . . . . . . .15 grs. Emmenagogue, parturient. See note 58.	9 00	4 75

No.		Per doz. pints.	Per gallon.
28	Orange flowers, U. S. P.	$ 7 50	$4 00
29	Orange peel, U. S. P.	5 75	3 75
30	Phosphates Comp., N. F. "Chem-		
	ical Food"	9 75	5 50
	Calcium phosphate . . . . 16 grs.		
	Iron phosphate . . . . . . 8 grs.		
	Potassium phosphate . : . . 3 grs.		
	Sodium phosphate . . . . 3 grs.		
	Ammonium phosphate . . . 8 grs.		
	Tonic.		
31	Rhatany, U. S. P.	8 50	4 75
	100 Cc. represent— 1 fl. oz. represents—		
	45 Cc. Fl. Ext. Rhatany . . 216 m.		
	Alterative, laxative.		
32	Rhubarb, U. S. P.	8 00	4 50
	100 Cc. contain— 1 fl. oz. contains—		
	10. Cc. Fl. Ext. Rhubarb. 48 m.		
	.4 Cc. Spirit Cinnamon. . 2 m.		
	1. Gm. Potass. carb. . . 4 1-2 grs.		
	Alkaline laxative, antidiarrheic.		
33	Rhubarb Aromatic, U. S. P. . .	7 20	3 80
	100 Cc. contain— 1 fl. oz. contains—		
	15 Cc. Arom. Tinct. Rhubarb. 68 m.		
	Gentle laxative; antidiarrheic.		
34	Rhubarb and Potassa Comp.		
	"Neutralizing Cordial"	9 00	5 00
	Rhubarb . . . . 30 grs.		
	Cinnamon . . . . 15 grs.		
	Golden Seal . . . . . . 10 grs.		
	Potassium carbonate . . . 3 1-2 grs.		
	Brandy . . . . q. s.		
	Oil of Peppermint . . . q. s.		
	Alkaline laxative, antidiarrheic.		
35	Sarsaparilla Comp., U. S. P. . .	8 50	4 75
	100 Cc. represent— 1 fl. oz. represents—		
	20. Cc. Fl. Ext. Sarsaparilla 196 m.		
	1.5 Cc. Fl. Ext. Senna . . . 7 m.		
	1.5 Cc. Fl. Ext. Glycyrrhiza 7 m.		
	Alterative.		
36	Sarsaparilla Comp. with Potassium		
	iodide	9 00	5 00
	Sarsaparilla . . . . 90 grs.		
	Senna . . . . 7 1-2 grs.		
	Glycyrrhiza . . . . . . 7 1-2 grs.		
	Potassium iodide . . . . 8 grs.		
	Alterative.		

No.		Per doz. pints.	Per gallon.
37	Senega, U. S. P.	$11 25	$6 50
	100 Cc. represent— 1 fl.oz. represents—		
	20. Cc. Fl. Ext. Senega . . . . 91 m.		
	Expectorant.		
38	Senna, U. S. P.	9 50	5 50
	100 Cc. represent— 1 fl.oz. represents—		
	25 Gm. Senna . . . . . 114 grs.		
	Laxative.		
39	Sodium hypophosphite, 32 grs.	8 25	4 50
	Tonic.		
	Squaw Vine. See Mitchella Comp.		
	(27)	6 00	4 75
40	Squill, U. S. P.	7 20	3 75
	100 Cc. contain— 1 fl. oz. contains—		
	45 Cc. Vinegar of Squill . . . 216 m.		
	Expectorant.		
41	Squill Comp., U. S. P.	9 00	4 75
	100 Cc. represent— 1 fl.oz. represents—		
	8. Cc. Fl. Ext. Squill . . . 38 m.		
	8. Cc. Fl. Ext. Senega . . . 38 m.		
	.2 Gm. Tartar Emetic. . . 9-10 gr.		
	Expectorant, emetic.		
42	Stillingia Comp.	9 00	4 75
	Stillingia . . . . 30 grs.		
	Corydalis. . . . . . . 30 grs.		
	Blue Flag . . . . . . 15 grs.		
	Elder flowers . . . . . 15 grs.		
	Chimaphila . . . . . . 15 grs.		
	Prickly ash. . . . . 7 1-2 grs.		
	Coriander . . . . . 7 1-2 grs.		
	Alterative.		
43	Tar, U. S. P.	6 50	3 50
44	Tolu, U. S. P.	7 00	3 75
	Tonic Hypophosphites. See		
	Hypophosphites, Tonic (16) .	9 00	5 00
45	Trifolium Comp.	11 25	7 00
	Red Clover . . . . . 32 grs.		
	Stillingia . . . . . . 16 grs.		
	Berberis Aquifolium. . . . 16 grs.		
	Burdock root . . . . . 16 grs.		
	Cascara Amarga . . . . 16 grs.		
	Poke root . . . . . . 16 grs.		
	Prickly ash bark . . . . . 4 grs.		
	Potassium iodide . . . . 8 grs.		
	Alterative. See note 101.		

No.		Per doz. pints.	Per gallon.
46	White Pine Comp. . . . . White Pine . . . . 30 grs. Wild Cherry . . . . 30 grs. Balm of Gilead buds . . . . . . 4 grs. Spikenard . . . . 4 grs. Blood-root . . . . 4 grs. Sassafras . . . . 2 grs. Chloroform . . . . 3 3-4 m. Morphine acetate . . . . . 2-10 gr. Expectorant. See note 104.	$ 7 50	$4 00
47	Wild Cherry, U. S. P. . . . . . . 100 Cc. represent— 1 fl. oz. represents— 15 Gm. Wild Cherry . . . . 68 grs. Sedative expectorant.	9 00	4 75
48	Yerba Santa Aromatic . . . . . . Yerba Santa . . . . 60 grs. Blackberry . . . . 30 grs. Licorice . . . . 15 grs. Aromatics . . . . q. s. Vehicle. See note 105.	10 00	6 00

CONCENTRATIONS.
(RESINOIDS)

In the following list we offer only such concentrations (or resinoids) as have been proven trustworthy remedial agents. The experience of the past few years has shown that of the many which were originally placed on the market but few possessed any therapeutic value, therefore we have revised our list carefully from time to time and have eliminated those which have not seemed to merit a place therein.

	Per oz.
Euonymin, green	$1 10
Hydrastin	1 50
Jalapin . . . . Resin of Jalap, U. S. P.	90
Leptandrin	40
Podophyllin (per lb. $5.00) . . . . Resin of Podophyllum, U. S. P.	40
Podophyllin, yellow (per lb. $5.00) . . .	40
Podophyllin, soluble, in scales (per lb. $6.00) . See note 91.	50

MEDICINAL WINES.

These are the wines of the U. S. Pharmacopœia with the addition of some others of established value. Unless otherwise specified the medicinal ingredients are stated in grains per fluidounce. The prices quoted are for pints of 16 fl. ozs. and for gallons in half-gallon bottles, no extra charge being made for containers.

No.		Per doz. pints.	Per gallon.
1	American Ash, 120 grs.	$9 75	$5 50
2	Antimony, U. S. P.	10 50	6 00
	100 Cc. contain: 1 fl. oz. contains: .4 Gm. Tartar Emetic. . . 1 4-5 grs.		
3	Aromatic.	12 00	7 00
4	Beef (Beef and Wine)	9 00	5 00
	Beef peptone . . . . 32 grs. See note 10.		
5	Beef and Iron (Beef, Iron and Wine)	9 00	5 00
	Beef peptone . . . . 32 grs. Peptonized Iron . . . . . 4 grs. See note 10.		
6	Beef, Iron and Cinchona (Beef, Iron, Cinchona and Wine) . . .	9 75	5 50
	Beef peptone . . . . . . 32 grs. Peptonized Iron . . . . . 4 grs. Red Cinchona . . . . . 10 grs. See note 10.		
7	Beef and Iron with Lactinated Pepsin (Beef, Iron and Wine with Lactinated Pepsin).	12 00	7 00
	Lactinated Pepsin . . . . 16 grs. Beef peptone . . . . . . 32 grs. Peptonized Iron . . . . . 4 grs. See notes 10, 71 a.		
8	Beef, Iron and Pepsin (Beef, Iron, Pepsin and Wine)	12 00	7 00
	Beef peptone . . . . 32 grs. Peptonized Iron . . . . . 4 grs. Saccharated Pepsin . . . . 16 grs. See note 10.		
9	Calisaya, 20 grs.	7 50	4 00
10	Coca, 60 grs. (See note 31.) . . .	9 75	5 50

No.	EDICINAL WINES.	Per doz. pints.	Per gallon.
11	Coca, aromatic. (See note 31). Coca . . . . 60 grs Aromatics . . . . q. s.	$ 9 75	$5 50
12	Coca and Beef (Coca, Beef and Wine). (See notes 10, 31) . . . Coca . . . . 60 grs. Beef peptone . . . . 32 grs.	10 50	6 00
13	Coca, Beef and Iron (Coca, Beef, Iron and Wine). . . . . Coca . . . . 60 grs. Beef peptone . . . . 32 grs. Peptonized Iron . . . . . . 4 grs. See notes 10, 31.	10 50	6 00
14	Colchicum root, U. S. P. . . . . 100 Cc. represent— 1 fl.oz.represents— 40 Gm. Colchicum root. . . 182 grs.	9 75	5 50
15	Colchicum seed, U. S. P. . . . . 100 Cc. represent— 1 fl.oz.represents— 75 Gm. Colchicum seed . . . 68 grs.	9 75	5 50
16	Ergot, U. S. P. . . . . 100 Cc. represent— 1 fl.oz.represents— 15 Gm. Ergot . . . . 68 grs.	12 75	7 50
17	Ipecac, U. S. P. . . . . 100 Cc. contain— 1 fl. oz. contains— 10 Gm. Fl. Ext. Ipecac. . . . 48 m.	16 50	10 00
18	Iron, B. P.	9 75	5 50
19	Iron, bitter, U. S. P. . . . . . 100 Cc. contain— 1 fl. oz. contains— 5 Gm. Iron and Quin. citrate 23 grs.	9 75	5 50
20	Iron, bitter, Parrish's. . . . . . Iron and Ammonium citrate . 8 grs. Extract Calisaya . . . . . . 1 gr.	9 75	5 00
21	Iron citrate, U. S. P. . . . . . . 100 Cc. contain— 1 fl. oz. contains— 4 Gm. Iron and Ammon. cit. 18 grs.	8 25	4 50
22	Opium. (See note 82) . . . . 100 Cc. represent— 1 fl.oz.represents— 10 Gm. Opium . . . . . . 46 grs.	22 50	14 60
23	Pepsin, saccharated, 40 grs. . . .	10 50	6 00
24	Tar	7 50	4 00
25	Wild Cherry, ferrated . . . . Wild Cherry . . . . 120 grs. Ferric pyrophosphate. . . . 4 grs.	9 75	5 50

LACTINATED PEPSIN AND ITS PREPARATIONS.

The combination of digestive ferments which we have designated as Lactinated Pepsin, and which is more fully described in Note 71 a, is one that has been found of considerable value in medicine. In the following list are named the various preparations of it that have come into general use, including certain combinations with other medicinal agents whose action is assisted by its influence on digestion and nutrition. Some of these preparations are quoted elsewhere under appropriate headings, but for the sake of convenience we present herewith the complete list.

Each fluidounce contains the specified quantity or medicinal ingredients.

No.		Per doz. pints.	Per gallon.
	Lactinated Pepsin (in powder), per oz., 50c.; per lb., $6.00. See note 71 a.		
7	Lactinated Pepsin with Beef, Iron and Wine	$12 00	$7 00
	Lactinated Pepsin . . . . . 16 grs.		
	Beef peptone . . . . 32 grs.		
	Peptonized iron . . . . 4 grs.		
	Sherry wine . . . . 1 oz.		
	See notes 10, 71 a.		
	Liquid Lactinated Pepsin	12 00	7 00
	Lactinated Pepsin . . . . 38 grs.		
	In a glycerin menstruum.		
	See note 71 a.		
22	Syrup of Lactinated Pepsin with Phosphates	12 00	7 00
	Lactinated Pepsin . . . . . 32 grs.		
	Iron phosphate . . . . 8 grs.		
	Calcium phosphate . . . . . 8 grs.		
	Sodium phosphate . . . . . . 8 grs.		
	Potassium phosphate . . . . . 8 grs.		
	See note 71 a.		

No.		Per doz. pints.	Per gallon.
81	Elixir Lactinated Pepsin, 38 grs. . . See note 71 a.	$12 00	$6 00

82	Elixir Lactinated Pepsin and Bismuth . . . .
	    Lactinated Pepsin . . . . . 32 grs.
	    Bismuth and Ammon. citrate. 8 grs.
	    See note 71 a.
		12 00	7 00

83	Elixir Lactinated Pepsin, Bismuth and Strychnine . . . .
	    Lactinated Pepsin . . . . . 32 grs.
	    Bismuth and Ammon. citrate. 8 grs.
	    Strychnine . . . . 1-12 gr.
	    See note 71 a.
		12 00	7 00

84	Elixir Lactinated Pepsin and Calisaya . . . .
	    Lactinated Pepsin . . . . . 32 grs.
	    Calisaya bark. . . . . . . 50 grs.
	    See note 71 a.
		12 00	7 00

85	Elixir Lactinated Pepsin, Calisaya, Iron and Bismuth . . . . . .
	    Lactinated Pepsin . . . . . 32 grs.
	    Calisaya bark . . . . . . 40 grs.
	    Ferric pyrophosphate . . . . 12 grs.
	    Bismuth and Ammon. citrate. 2 grs.
	    See note 71 a.
		12 00	7 00

86	Elixir Lactinated Pepsin, Gentian and Chloride of Iron . . . . .
	    Lactinated Pepsin . . . . . 32 grs.
	    Gentian . . . . 8 grs.
	    Proto-chloride of Iron . . . . 8 grs.
	    See note 71 a.
		12 00	7 00

87	Elixir Lactinated Pepsin with Iron, Quinine and Strychnine phosphates . . . .
	    Lactinated Pepsin . . . . . 32 grs.
	    Ferric phosphate . . . . . 4 grs.
	    Quinine phosphate. . . . . 4 grs.
	    Strychnine phosphate . . . 1-16 gr.
	    See note 71 a.
		12 00	7 00

TINCTURES OF THE U. S. PHAR-MACOPŒIA AND NATIONAL FORMULARY.

The following is a complete list of the tinctures of the U. S. Pharmacopœia and National Formulary, with a few unimportant exceptions. All that are not referred to any authority are those of the pharmacopœia, and such as are of determinable strength are assayed to conform to definite standards.

No.		Per ℔.
1.	Aconite root, assayed	$1 00
	Standard: 1:250 dilution, physiological test.	
2.	Aconite root, N. F., Fleming's, assayed .	1 00
	Standard: 1:500 dilntion, physiological test.	
3.	Aloes	85
4.	Aloes and Myrrh	1 00
5.	Antacrid, N. F.	1 25
	Antiperiodic, N. F. See Warburg's . .	3 00
6.	Arnica flowers	90
7.	Arnica root, ...	90
8.	Aromatic, N. F.	90
9.	Asafetida	1 00
10.	Belladonna leaves, assayed	1 00
	Standard: .06% alkaloids, by titration.	
11.	Benzoin	1 00
12.	Benzoin Comp.	1 00
13.	Bitter, N. F.	1 00
14.	Blood-root, assayed	1 00
	Standard: .5% alkaloids, by titration.	
15.	Bryonia	1 00
16.	Calendula	1 00
17.	Cannabis Indica	1 00
18.	Cantharides	1 00
19.	Capsicum	1 00
20.	Capsicum and Myrrh, N. F.	1 00
21.	Cardamom	1 25
22.	Cardamom Comp.	1 00
23.	Catechu Comp.	1 00
24.	Chirata	1 00

TINCTURES OF THE U. S. PHARMACOPŒIA AND N. F.—Continued

		Per ℔
25.	Cimicifuga	$1 00
26.	Cinchona, assayed	1 00
	Standard: 5% total alkaloids, by weight.	
27.	Cinchona Comp., assayed	1 00
	Standard: .5% total alkaloids, by weight.	
28.	Cinchona, detannated, N. F. [Phar]	1 00
29.	Cinnamon	1 00
30.	Colchicum seed, assayed	1 00
	Standard: .07% colchicine, by weight.	
31.	Columbo	1 00
32.	Conium, N. F., assayed	1 00
	Standard: .07% coniine, by titration.	
33.	Coto, N. F.	1 50
	See note 36.	
34.	Cramp-bark Comp., N. F.	1 00
35.	Cubeb	1 00
36.	Cudbear, N. F.	1 00
37.	Cudbear Comp., N. F.	1 00
38.	Digitalis	1 00
39.	Galls	1 00
40.	Gelsemium, assayed	1 00
	Standard: .07% total alkaloids, by titration.	
41.	Gentian Comp.	1 00
42.	Ginger	1 00
43.	Guaiac	1 00
44.	Guaiac, ammoniated	1 25
45.	Guaiac Comp., N. F.	1 00
46.	Henbane, assayed	1 00
	Standard: .02% total alkaloids.	
47.	Hops	1 00
48.	Hydrastis, assayed	1 00
	Standard: .5% total alkaloids, by titration.	
49.	Ignatia, N. F., assayed	1 00
	Standard: .2% total alkaloids.	
50.	Iodine	1 50
51.	Iodine, Churchill's, N. F.	2 50
52.	Iodine, colorless, N. F.	1 50
53.	Ipecac and Opium	2 40
54.	Iron chloride	75
55.	Iron chloride, ethereal, N. F.	1 00
56.	Iron citro-chloride, N. F.	1 00

No.		Per lb.
57.	Jalap, N. F.	$1 00
58.	Jalap Comp., N. F.	1 00
59.	Kino	2 00
60.	Kino Comp., N. F.	1 50
61.	Lactucarium	3 50
62.	Lavender Comp.	1 00
63.	Lobelia	1 00
64.	Matico	1 00
65.	Musk-root	1 00
66.	Myrrh	1 00
67.	Nux Vomica, assayed	1 00
	Standard: .3% total alkaloids, by titration.	
68.	Opium, assayed	1 75
	Standard: 1.4% morphine.	
69.	Opium, camphorated	1 00
70.	Opium, deodorized	2 50
	Standard: 1.4% morphine.	
71.	Orange peel, bitter	1 00
72.	Orange peel, sweet	1 00
73.	Pectoral, N. F.	1 00
74.	Pellitory	1 00
75.	Physostigma, assayed	1 00
	Standard: .04% physostigmine, by weight.	
76.	Pimpinella, N. F.	1 00
77.	Poppy, N. F.	1 00
78.	Quassia	1 00
79.	Rhatany	1 00
80.	Rhubarb	1 00
81.	Rhubarb, aqueous, N. F.	1 00
82.	Rhubarb, aromatic	1 00
83.	Rhubarb and Gentian, N. F.	1 00
84.	Rhubarb, sweet	1 25
85.	Rhubarb, vinous, N. F.	1 00
86.	Serpentaria	1 00
87.	Soap-tree bark	85
88.	Soap, green, Comp., N. F.	1 00
89.	Squill	85
90.	Stramonium seed, assayed	1 00
	Standard: .05% total alkaloids, by weight.	
91.	Strophanthus	1 50

No.		Per ℔.
92.	Tolu.	$1 00
93.	Tolu, soluble, N. F.	1 00
94.	Valerian	1 00
95.	Valerian, ammoniated	1 25
96.	Vanilla, U. S. P.	2 00
97.	Vanillin Comp., N. F.	2 00
98.	Veratrum Viride, assayed	1.00
	Standard: .4% total alkaloids, by weight.	
99.	Warburg's, with Aloes, N. F.	3 00
100.	Warburg's, without Aloes, N. F.	3 00
101.	Zedoary Comp., N. F.	1 50

TINCTURES OF THE BRITISH PHARMACOPŒIA.

For the convenience of our patrons that use the British Pharmacopœia as the standard of authority, we present a separate list of the tinctures official in that publication, and which are prepared according to the requirements of its latest revision.

No.		Per ℔.
102.	Aconite root	$1 00
103.	Aloes	85
104.	Arnica root	90
105.	Asafetida	1 00
106.	Belladonna, assayed	1 00
	Standard: .05% alkaloids.	
107.	Benzoin Comp.	1 00
108.	Buchu	1 00
109.	Calumba	1 00
110.	Camphor Comp.	1 00
111.	Cannabis Indica	1 00
112.	Cantharides	1 00
113.	Capsicum	1 00
114.	Cardamom Comp.	1 00
115.	Cascarilla	1 00
116.	Catechu	1 00
117.	Chirata	1 00

No.		Per ℔.
118.	Chloroform and Morphine Comp.	$1 00
119.	Cimicifuga	1 00
120.	Cinchona, assayed	1 00
	Standard: 1.% alkaloids.	
121.	Cinchona Comp., assayed	1 00
	Standard: .5% alkaloids.	
122.	Cinnamon	1 00
123.	Cochineal.	1 00
124.	Colchicum seed	1 00
125.	Conium	1 00
126.	Cubeb	1 00
127.	Digitalis	1 00
128.	Ergot, ammoniated	1 00
129.	Gelsemium	1 00
130.	Gentian Comp.	1 00
131.	Ginger	1 00
132.	Guaiac, ammoniated	1 25
133.	Hamamelis	1 00
134.	Henbane	1 00
135.	Hops	1 00
136.	Hydrastis	1 00
137.	Iodine	1 50
138.	Iron chloride	75
139.	Jaborandi	1 00
140.	Jalap	1 00
141.	Kino	2 00
142.	Lavender Comp.	1 00
143.	Lemon peel	1 00
144.	Myrrh	1 00
145.	Nux Vomica, assayed	1 00
	Standard: .25% strychnine.	
146.	Opium, assayed	1 75
	Standard: .75% anhydrous morphine.	
147.	Opium, ammoniated	1 25
	Opium, camphorated. See Camphor Comp. (110)	1 00
148.	Orange peel, bitter	1 00
149.	Podophyllum	1 00
150.	Pyrethrum	1 00
151.	Quassia	1 00

No.		Per ℔.
152.	Quillaia	$ 85
153.	Quinine	90
154.	Quinine, ammoniated	90
155.	Rhatany	1 00
156.	Rhubarb Comp.	1 00
157.	Senega	1 00
158.	Senna Comp.	1 00
159.	Serpentaria	1 00
160.	Squill	85
161.	Stramonium seed	1 00
162.	Strophanthus	1 50
163.	Sumbul	1 00
164.	Tolu	1 00
165.	Valerian, ammoniated	1 25
166.	Wild Cherry	1 00

SOLUTIONS.

Per ℔.

Acid Phosphates. See Phosphoric Acid Comp.
Per gal., $1.75; per doz. pts., $5.00; per
doz. half-pints, $2.75.

Ammonia, 10% (Water of Ammonia) .... $ 25
Per 5 pint g. s. bottle, incl., $1.05.

Ammonia, 26% (Stronger Water of Ammonia) . 35
Per 5 pint g. s. bottle, incl., $1.40.

Ammonium acetate, U. S. P. 80

Arsenic, Fowler's. See Potassium Arsenite. 35

Arsenic and Mercuric iodide, U. S. P. . . . 50

Arsenous acid, U. S. P. 55

Atropine sulphate, B. P., per oz. 30c.

Cocaine hydrochlor., 2.%, per oz., 30c.

Cocaine hydrochlor., 4%, per oz., 50c.

Corrosive, P. G. 1 00
Donovan's. See Arsenic and Mercuric iodide. 50
Fowler's. See Potassium arsenite. 35
Goulard's. See Lead subacetate . . . 1 40
Hall's. See Strychnine acetate 80

	Per ℔.
Iodine Comp.	$2 00
Iron acetate, U. S. P.	1 10
Iron acetate, P. G.	1 10
Iron chloride, U. S. P.	40
Iron citrate, U. S. P.	1 10
Iron iodide, Conc., for syrup	7 50
Eight times as strong as the U. S. P. syrup.	
Iron nitrate, U. S. P.	60
Iron subsulphate, U. S. P.	40
Basic Ferric Sulphate.	
Iron tersulphate, U. S. P.	45
Normal Ferric Sulphate.	
Javelle's. See Potassa, chlorinated.	30
Labarraque's. See Soda, chlorinated	30
Lead subacetate, U. S. P.	40
Lead subacetate, diluted, U. S. P.	30
Lead water.	
Lime, saccharated, B. P.	75
Lugol's. See Iodine Comp.	2 00
Mercuric nitrate, U. S. P., per oz., 30c.	
Mercury perchloride, B. P.	35
Monsel's. See Iron subsulphate	40
Pearson's. See Sodium arsenate	80
Pepsin, N. F.	1 40
Phosphoric acid, Compound, (Acid Phosphates), per gal., $1.75; per doz. pints, $5.00; per doz. half pints, $2.75.	
See note 1.	
Potassa, U. S. P. (Potassium hydrate)	35
Potassa, chlorinated, N. F.	30
Potassium arsenite, U. S. P.	35
Potassium citrate, U. S. P.	1 40
Potassium permanganate, B. P.	40
Soda, chlorinated, U. S. P.	30
Sodium arsenate, N. F.	80
Sodium silicate, U. S. P. (Water Glass).	30
Strychnine acetate, N. F.	80
Strychnine hydrochlorate, B. P.	80
Zinc chloride, U. S. P.	50

INHALANTS.

For use in atomizers.

| | | Per lb. |

No. 1 Inhalant. **$1 00**
  Copaiba . . . . . . q. 1-2 oz.
  Ether . . . . 1 oz.
  Liquid Puraline . . . q. s. to 4 ozs.

No. 2 Inhalant. **1 25**
  Fl. Ext. Stramonium . . . . 1 dr.
  Fl. Ext. Hyoscyamus . . . 1 dr.
  Fl. Ext. Belladonna . . . 1-2 dr.
  Glycerin . . . . 1 oz.
  Alcohol . . . . . . q. s. to 3 ozs.

No. 3 Inhalant. **1 00**
  Comp. Tincture Benzoin . . 1 oz.
  Glycerin . . . . . . 1 oz.
  Alcohol . . . . 1 oz.

No. 4 Inhalant. **60**
  Oil Tar . . . . 1-2 dr.
  Liquid Puraline . . . . . 1 oz.

No. 5 Inhalant. **1 00**
  Oil Eucalyptus . . . . . 1 dr.
  Liquid Puraline . . . . . 1 oz.

No. 6 Inhalant. **1 00**
  Tincture Iodine . . . . 1 1-2 drs.
  Glycerin . . . . 1 oz.
  Alcohol . . . . . . q. s. to 3 ozs.

No. 7 Inhalant. **1 00**
  Tincture Iodine . . . . . 2 drs.
  Carbolic acid . . . . . 3 drs.
  Fl. Tolu, soluble . . . . 1 oz.
  Glycerin . . . . 1 oz.
  Alcohol . . . . . q. s. to 3 ozs.

No. 8 Inhalant. **1 00**
  Tincture Iodine . . . . . 1-2 dr.
  Glycerin . . . . . . 1-2 oz.

No. 9 Inhalant. **60**
  Carbolic acid . . . . . 10 grs.
  Liquid Puraline . . . . . 1 oz.

No. 10 Inhalant **1 00**
  Each fl. oz. represents 1-2 ounce
  Compound Tincture Benzoin.

No. 11 Inhalant *discontinued*.

No. 12 Inhalant **1 00**
  Boroglyceride (50.%) . . . . 1 oz.
  Glycerin . . . . . . 1 oz.
  Ether . . . . 2 ozs.

OLEATES.

These preparations are in most cases true chemical combinations either with or without an excess of oleic acid. The metallic oleates are waxy solids or unctuous powders and may be diluted to any desired extent. The "oleates of the alkaloids" are oleic acid solutions of the true "oleates." The facility with which oleates are absorbed by the skin renders them an important and valuable class of medicaments. For further information see Note 81.

A list of ointments of oleates is also appended.

OLEATES OF ALKALOIDS.
See note 81.

No.		Per oz.
1.	Aconitine, crystalline	$ 80
	Containing 2.% alkaloid.	
2.	Atropine	65
	Containing 2.% alkaloid.	
3.	Cocaine	1 50
	Containing 5.% alkaloid.	
4.	Hydrastine	1 25
	Containing 2.% alkaloid.	
5.	Morphine	1 15
	Containing 10.% alkaloid.	
6.	Morphine and Mercury, "A"	70
	Containing { Mercury . . . 10.% { Morphine . . 5.%	
7.	Morphine and Mercury, "B".	75
	Containing { Mercury . . . 20.% { Morphine . . 5.%	
8.	Quinine	50
	Containing 25.% alkaloid.	
9.	Strychnine	50
	Containing 2.% alkaloid.	
10.	Veratrine, "A," U. S. P.	75
	Containing 2.% alkaloid.	
11.	Veratrine, "B".	1 00
	Containing 10.% alkaloid.	

OLEATES OF METALS.
See note 81.

No.		Per oz.	Per ℔.
12.	Arsenic	$ .30	$3 00
	Used in ointment form as a caustic.		
13.	Bismuth	35	4 50
	Used, undiluted, as an emollient.		
14.	Copper	25	3 00
	Used as an ointment for ringworm.		
15.	Lead	30	3 00
	An efficient substitute for Lead Plaster, U. S. P.		
16.	Manganese	50	5 00
	Recommended as an efficient emmenagogue by Drs. Mc-Arthur and Martin. Generally used in 20% ointment, and applied by inunction to the abdomen.		
17.	Mercury	35	4 50
	For producing the full therapeutic effects of mercury.		
18.	Nickel	60	6 50
	Reputed of value in skin affections.		
19.	Silver	2 50	25 00
20.	Tin	35	4 50
	Has been recommended for mani-cure purposes, especially for tinting and polishing the nails.		
21.	Zinc	25	3 00
	Used as a dusting powder in skin diseases, on account of its mildly astringent properties. Also employed in ointment form.		

OINTMENTS OF OLEATES.

		Per ℔.
22.	Copper oleate, 20.%	$1 80
23.	Copper oleate, 10.%	1 50
24.	Lead oleate, 50.%	2 50
25.	Lead oleate, 25.%	1 75
26.	Manganese oleate, 20.%	3 75
27.	Mercury oleate, 20.%	2 50
28.	Mercury oleate, 10.%	1 75
29.	Silver oleate, 5.%	2 75
30.	Zinc oleate, 25.%	1 75

·OINTMENTS AND CERATES.

· 'These ointments are made with a saxoline (petrolatum) base, never become rancid, and are perfectly smooth and homogeneous. They are put up in handsome, air-tight glass jars, for which no extra charge is made. For further information see Note 80.

This list also includes the official cerates.

No.		Per ℔. jar.
1.	Ammoniated Mercury, U. S. P.	$1 35
2.	Arnica flowers.	1 20
	Basilicon. See Resin (30)	75
3.	Belladonna, U. S. P.	1 15
	Blistering. See Cantharides (6)	1 25
4.	Boric acid, B. P.	1 15
5.	Camphor, U. S. P. (Cerate)	90
6.	Cantharides, U. S. P. (Cerate)	1 25
7.	Carbolic acid, U. S. P.	90
8.	Carbolic acid, improved	1 15

Carbolic acid. . . . . . . 13 parts.s
Mild Mercurous chloride. . 4 parts.
Camphor . . . . 1 part.
Ointment . . . . 208 parts.

9.	Chrysarobin, U. S. P.	1 75
	Citrine. See Mercuric nitrate (24)	1 00
	Cold Cream. See Rose Water (31)	1 50
10.	Diachylon, U. S. P.	1 15
11.	Gallic acid	1 15
12.	Galls, U. S. P.	90
13.	Galls and Opium, B. P.	2 25
14.	Goa powder, 2.5%	1 75
15.	Iodine, U. S. P.	1 25
16.	Iodoform, U. S. P.	1 75
17.	Itch	1 20

Potassium carbonate . . . 8 parts.
Sodium sulphite . . . . 16 parts.
Carbolic acid . . . . . . 1 part.
White Hellebore . . . . . 4 parts.
Ointment . . . . 64 parts.
Water . . . . 4 parts.

18.	Lead carbonate, U. S. P.	90
19.	Lead iodide, U. S. P.	1 90
20.	Lead subacetate, U. S. P. (Cerate).	1 00

No.		Per ℔. jar.
21.	Mayer's	$1 15
	Olive oil . . . . 10 parts.	
	White turpentine . . . . . 2 parts.	
	Red Lead . . . . 2 parts.	
	Butter . . . . 1 part.	
	Beeswax . . . . 1 part.	
	Camphor . . . . 1 part.	
	Honey . . . . 1 1-2 parts.	
22.	Mercurial, U. S. P., 50.% mercury	1 10
23.	Mercurial, 33.% mercury	60
24.	Mercuric nitrate, U. S. P.	1 00
25.	Mercuric oxide, red, U. S. P.	1 20
26.	Mercuric oxide, yellow, U. S. P.	1 50
27.	Mezereon.	90
28.	Pile	1 75
	Galls . . . . 4 parts.	
	Carbolic acid . . . . 4 parts.	
	Extract Ergot. . . . . . . 2 parts.	
	Extract Stramonium . . . 3 parts.	
	Extract Witch Hazel . . . 4 parts.	
	Oil Tar . . . . 2 parts.	
	Ointment . . . . 40 parts.	
29.	Potassium iodide, U. S. P.	1 50
30.	Resin, U. S. P. (Cerate)	75
31.	Rose Water, U. S. P.	1 50
32.	Savine	1 20
33.	Salt Rheum.	1 20
	Ammoniated Mercury . . . 3 parts.	
	Lead acetate . . . . 3 parts.	
	Carbolic acid . . . . 2 parts.	
	Ointment . . . . 45 parts.	
34.	Simple, U. S. P.	75
35.	Stramonium, U. S. P.	90
36.	Sulphur, U. S. P.	75
37.	Sulphur, Alkaline, N. F.	90
38.	Sulphur Comp., N. F.	1 00
39.	Tannic acid, U. S. P.	1 35
40.	Tar, U. S. P.	75
41.	Tar Compound	1 00
	Tar . . . . 25 parts.	
	Sulphur . . . . 15 parts.	
	Ointment . . . . 60 parts.	
42.	Zinc oxide, U. S. P.	**75**

Prices given for Biologic Products, in Part II of this
Catalogue, are subject to the stated discounts.
An additional discount of 5% is conceded
on lots of not less than $25 net.

PART SECOND.

BIOLOGIC PRODUCTS

AND

SPECIALTIES.

Prices given for the Specialties, in Part II of this
Catalogue, are strictly Net, except in quantity
lots, for which special prices will
be made on application.

STEARNS' VACCINE.

The Kind That Vaccinates Without Complications.

Bacteriologically and Physiologically Tested and Perfectly Correct in Every Particular.

Stearns' Vaccine represents the highest attainments in vaccine production, and has no superior. In its elaboration only healthy and robust calves are used and every precaution is exercised to have the product free from septic contamination. The most modern and improved procedures are used in its production.

In primary vaccinations, Stearns' Vaccine will produce the typical vesicle in 100 per cent. of cases, while in secondary cases, it will "take," if any vaccine will. If properly applied it will not cause excessive inflammation and tumefaction.

GLYCERINATED—IN SEALED TUBES.

It is uniquely put up, each package comprising a complete vaccination.

IN DISPENSING PACKAGE (10 tubes, each tube in separate sterilized package containing also an ejector, scarifying needle and temporary shield) $1 00
Discount 20%.

IN STOCK PACKAGE (10 tubes in one wooden tube, accompanied by one needle, rubber bulb and ten temporary shields) $1 00
Discount 50%.

IVORY POINTS—NOT GLYCERINATED.

We furnish non-glycerinated vaccine, of exceptional purity, on ivory points. The points are prepared under strictest aseptic precautions and every care is taken to guard them against septic contamination. Each one is carefully wrapped in a separate sterilized paper container, and is thus protected until ready for use.

Per packages of 10 . . . . $1 00
Discount 50%.

STEARNS' ANTISTREPTOCOCCIC SERUM.

It is supplied in the same convenient device that has rendered Stearns' "Special" Diphtheritic Antitoxin so popular. It is always ready for use, and requires no extra syringe, since it can be injected directly from its own bulb, thus assuring absolute asepsis.

Each package contains 20 c. c. of the serum, in two bulbs of 10 c. c. each, so that it may readily be administered in either prophylactic or curative doses.

Per package. $1 75
Discount 30%.

STEARNS' SPECIAL DIPHTHERITIC ANTITOXIN
Distinguished by its Many Advantages.

THE PACKAGE

1. Is a unique hypodermic syringe when ready for use, by which the antitoxin may be injected directly from its own bulb. 2. Furnishes a new syringe for each case. 3. Is perfectly aseptic. 4. Is convenient. Is made ready in a minute. No loss of time from sterilization, as with the older methods. 5. Is economical. 6. In every way far in advance of the old-time, clumsy, vexing and uncertain methods.

THE ANTITOXIN

1. Is produced by the most modern and improved methods. 2. Is carefully filtered. 3. Is bacteriologically tested. 4. Is tested and standardized on animals. 5. Each bulb is guaranteed to contain the number of units stated on the label, and to be correct in every particular. 6. Gives largest percentage of recoveries.

There is none better.

SPECIAL

In hypodermic package, as per illustration (Standard, 500, or more, units per Cc.)

No. 1—Syringe bulb of 500 units. . . .$1 15
No. 2—Syringe bulb of 1000 units. . . . 2 25
No. 3—Syringe bulb of 1500 units. . . . 3 25
No. 4—Syringe bulb of 2000 units. . . . 4 00
No. 5—Syringe bulb of 3000 units. . . . 5 75

REGULAR, OR STANDARD

In conventional flasks. Never offered in the syringe bulb. (Standard, 250 to 500 units per Cc.)

No. 1—Bulb of 500 units. $0 75
No. 2—Bulb of 1000 units. 1 50
No. 3—Bulb of 1500 units. 2 25
No. 4—Bulb of 2000 units. 3 00
No. 5—Bulb of 3000 units. 4 50

Discount 30%.

KASAGRA.

The original and only genuine Cascara Aromatic.
A fluid extract of prime Cascara Sagrada, aromatized
and sweetened, thus being rendered extremely palatable.

Kasagra contains *all* the active principles of the best selected bark of Rhamnus Purshiana, and owes its laxative properties to this drug *alone*.

It *does not gripe*, as does the ordinary U. S. P. fluid extract. In preparing Kasagra bark two years old or older is used, which lessens its griping tendencies and improves its taste.

DOSE: The best results are obtained by administering Kasagra *in small doses* (15 to 60 minims according to whether tonic or laxative effects are desired) well diluted with water, four times a day, half an hour before each meal, and again at bedtime.

	Per pint., net.
1-pint bottles	$1 20
5-pint bottles	1 10

STEARNS' WINE OF COD LIVER OIL WITH PEPTONATE OF IRON.

Stearns' Wine of Cod Liver Oil with Peptonate of Iron is an original and elegant preparation, which contains the alkaloids and curative active principles to be found in one-fourth its volume of pure Cod Liver Oil, as well as four grains of Peptonate of Iron to each fluidounce. This preparation is entirely free from the oily or fatty matter of Cod Liver Oil, and is therefore pleasant to the taste, thus making it a valuable compound wherever Cod Liver Oil and Iron are indicated.

Modern investigation has proved that the value of Cod Liver Oil as a medicinal agent is due not simply to the fact of its being an oil, but to the valuable active principles which it contains.

Per doz., net.

Pint bottles $8 00

TRITIPALM.

A highly concentrated compound fluid extract of fresh Saw Palmetto and Triticum, of which each

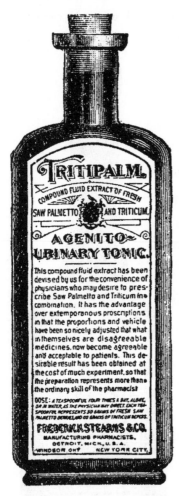

fluidrachm represents 30 grains of fresh Saw Palmetto berries and 60 grains of Triticum repens.

This compound fluid extract has been devised by us for the convenience of physicians who may desire to prescribe Saw Palmetto and Triticum in combination. It has an advantage over extemporaneous prescriptions in that the proportions and vehicle have been so nicely adjusted that what in themselves are disagreeable medicines, now become agreeable and acceptable to patients.

DOSE.—The dose of Tritipalm is a teaspoonful, taken alone or mixed with water, as the case may demand. The dose may be repeated four times a day, or as the exigencies may require.

Per doz., net.

8-oz. bottles $8 00

LIQUID HÆMOFERRUM.

A delicious cordial, each teaspoonful of which contains six grains of Hæmoferrum.

Hæmoferrum represents in a natural condition all the necessary constituents of the healthy, red-blood corpuscle, and in their normal proportions. It is nature's own compound for supplying the blood its characteristic oxygen-carrying power.

Hæmoferrum contains from 0.41 to 0.45 per cent. of iron. A small amount of absorbed iron is far more beneficial to the blood and system than large quantities of iron which traverse the alimentary tract in a form *useless* to the system. The entire amount of iron in the blood of a man weighing 145 pounds is only about 45 grains. Then, why give enormous doses of iron to make up a deficiency in this amount?

Per doz., net.

8-oz. bottles $8 00

VIBUTERO.

A Palatable and Effective Combination of the Two Viburnums with
Saw Palmetto and other Valuable Adjuvants.

A TRUE UTERINE TONIC.

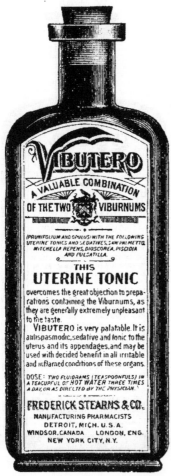

The idea of combining synergetic uterine tonics and restoratives in order to obtain a more effective action and a wider field of usefulness is not a new one. A number of such combinations have been before the profession for years. These possess more or less merit and have a large and growing demand. Yet there has been a chance for improvement, as every physician conversant with such preparations will readily admit. In appreciation of this fact and realizing the opportunity, a thorough study of the subject has been made, including much experimentation, and Vibutero is the result.

Composition:—Black Haw, Cramp-bark, Squaw-vine, Wild Yam, Jamaica Dogwood, Saw Palmetto berries and Pulsatilla, combined with suitable aromatics, making this the most palatable as well as the most potent of all preparations of its class.

8-oz. bottles . . . . $8 00 per dozen.

KOLA-STEARNS.

Kola-Stearns, a special aromatized fluid extract of fresh (undried), true African Kola, being its *only* palatable liquid preparation in concentrated form.

Kola-Stearns is a fanciful title which we have adopted to distinguish our brand of special fluid extract of Kola from other fluid extracts of Kola which may appear on the market. Being made from fresh (undried) Kola nuts, it possesses, undiminished, the same properties ascribed by travelers and by the natives in Africa to the fresh Kola nuts. Each dose ($\frac{1}{2}$ teaspoonful) represents 30 grains of the fresh (undried) Kola nut.

Kola is highly recommended by eminent authorities as a valuable stimulant to the brain, spinal cord and muscular system.

Per doz., net.

6-oz. bottles (96 doses) $8 00

DIASTASE-STEARNS.

Diastase is a name given to a number of substances, derived from various sources, that have the power of converting starches into sugars. There are a number of diastasic prepara-tions on the market made from malted grain, and although these transform starches in test-tube experiments, the question arises: "Can their action replace that of the natural pancreatic diastase?" Upon thoughtful consideration, there is some doubt that they can. The structure, life and needs of a plant are so different from those of the animal the inference is that a peculiarly adapted dias-tasic ferment would be applied to each. If the ultimate result of the action of diastase was alone to be considered it would not be a mat-ter of much moment as to what diastase to use. But the diges-tion of starches is marked by a series of different steps and the intermediate products so formed are variant in character. This gradual transformation is not without its usefulness. Nature never does anything haphazard. Does it not then stand to reason that the intermediate products formed by the action of an animal diastase are better suited to the animal economy than those so formed by a vegetable diastase?

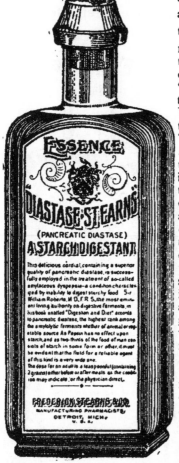

Diastase-Stearns is made from the fresh, healthy pancreas of the pig and is ten times stronger than vegetable diastase obtained from malt. It is pure, sweet and concentrated and is far superior for use as a medicine to the thick, molasses-like malt preparations or any dry diastase de-rivable from malt.

We offer Essence of Diastase-Stearns, a cordial containing in each teaspoonful two grains of Diastase-Stearns.

Per doz., net.

8-oz. bottles. $6 00

Diastase-Stearns is also marketed in powdered form at the following prices: ¼-oz. vial, 45 cents; ½-oz. vial, 80 cents; 1-oz. vial, $1.50.

PANZYME.

A new combination of Diastase-Stearns with other digestive ferments for treatment of indigestion and allied disorders.

In Panzyme, we have effected a happy union of digestive principles which differentiates it from all similar compounds and, when its virtues are properly appreciated by the practician, will place it easily pre-eminent in its class of digestive remedial agents.

Other preparations are generally presented singly and will act upon only one class of food, or if combined, although sometimes theoretically correct in physiologic proportions, are generally not satisfactory in results since malt or vegetable diastase (a substance foreign to the animal body) is used to supply the deficiency of the true pancreatic diastase. Another reason of their inefficiency is, the proportions of ingredients are generally too small since the alimentary canal requires larger amounts of digestants introduced from without than of those secreted in the system in a nascent state and just at the moment when needed.

We have given the subject a large amount of chemical, physiologic and clinical research and devoted much time to the careful study of digestants and their relations to each other and of their various combinations, before fully deciding to manufacture and offer to the profession our Panzyme, which is the culmination of our work in this line, and is a preparation of which we are justly proud.

Panzyme is not a secret-nostrum proprietary. It is composed of Diastase-Stearns, pancreatin, pepsin and rennin, combined in proper proportions and represents the entire digestive functions of the alimentary tract. These constituents are of the highest and purest quality, and are presented in the most easily soluble and most active condition. To these digestive enzymes are added grateful carminatives and aromatics.

Per dozen.

Vials of 50 tablets. $4 00

ZYMOLE.

Stearns' Antiseptic Compound.

The Ideal Antiseptic.

Composed of the antiseptic constituents of Gaultheria, Thyme, Eucalyptus, Baptisia, and Mentha arvensis, with the addition of Boric and Benzoic acids, Acetanilid, Formaldehyde and Boroglyceride.

Zymole is unsurpassed as a non-toxic, non-irritant and reliable antiseptic and deodorant, and its uses are manifold.

In 6-oz. special diamond bottles, per doz., net . $2 50
In bulk per pint, net 40
In bulk per 5 pints, net 1 90
In bulk per gallon, net 3 00

CAPSOIDS.

Filled capsules, without air, each containing 5 minims.

Per doz.
Flasks of 50 Capsoids.

Apiol $8 00
Apiol Compound 6 00
        Apiol . . . . 2 minims.
        Oil Savin . . . . 1 1-2 minims.
        Oil Tansy . . . . 1 1-2 minims.
Chlorodyne, Conc. 4 50
        Representing 15 minims Chlorodyne.
Copaiba Balsam 3 00
Copaiba and Santal (Each 2 1-2 minims) . . . 4 50
Creosote and Cod Liver Oil 4 00
        Creosote . . . . 2 minims.
        Cod Liver Oil . . . . . . 3 minims.
Cubeb Oil 6 00
Cubeb Oleoresin 6 00
Cubeb and Copaiba 4 50
        Cubeb Oleoresin . . . . . 2 minims.
        Copaiba Balsam . . . . . 3 minims.
Eucalyptus Oil 3 00
Guaiacol 3 00
        Guaiacol . . . . 1 minim.
        Sweet Almond Oil . . . . 4 minims.
Jecorol (Ext. Olei Morrhuæ, Stearns') 5 00
Jecorol and Creosote 5 00
        Jecorol, 3 minims; Creosote, 2 minims.
Male Fern Oleoresin 6 00
Salol and Copaiba (Each 2 1-2 grains) 4 50
Salol and Santal Oil (Each 2 1-2 grains) . . . 6 00
Santal Oil 5 00
Santal Compound 5 00
        Santal Oil . . . . 2 minims.
        Copaiba Oil . . . . . . 2 minims.
        Copaiba Oil . . . . . . 1 minim.
Saw Palmetto (Oleoresin) 5 00
Saw Palmetto Compound 6 00
        Oleoresin Saw Palmetto . . 3 minims.
        Santal Oil . . . . 2 minims.
Saw Palmetto Compound with Salol 6 00
Turpentine Oil 3 00
Wintergreen Oil (natural) 5 00
Wintergreen Oil and Salol (Each 2 1-2 grains). 5 00

COLSALOIDS.

Capsoids, each containing 1-250 grain of Colchicine in 3 minims of Methyl Salicylate.

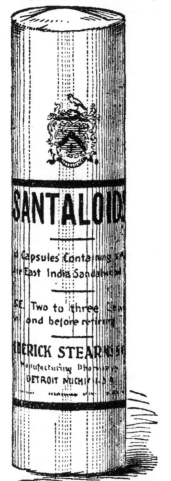

Colchicum is universally conceded, and has been for years, unequalled as an antarthritic and eliminative in cases of gout and gouty diathesis. It is a potent gastro-intestinal irritant, cholagogue and eliminative, highly accelerating the removal from the system of uric acid and xanthine, relieving and preventing toxemic conditions.

In Colsaloids is used the active principle of Colchicum, its alkaloid Colchicine, which, although small in amount, produces the characteristic action of the drug from which it is obtained.

Per doz., net.

In amber, wide-mouth
  vials of 50 each . . . $5 00

SANTALOIDS.

Capsoids, each containing 5 minims of Pure East India Sandalwood Oil.

The Santal Oil used in the preparation of this product is of the finest obtainable, responding completely to the very thorough test of the U. S. P.   Per doz., net.

In amber, wide-mouth vials of 40 each $4 50

SANTALOIDS COMP.

Capsoids, each containing 1 minim of Pure East India Sandalwood Oil, 1 minim of Copaiba Oil, and 3 minims of Haarlem Oil.   Per doz., net.

In amber, wide-mouth bottles of 40 each . . . $3 00

METHYLOIDS.
TRADE MARK.

An Improved Combination in Capsoid Form for the Successful Treatment of Gonorrhea, Its Complications, and all cases where a Urinary Antiseptic is Indicated.

Each Methyloid contains Methylene Blue, C. P., gr. 1; Santal Oil and Copaiba, of each, m. 1½; Haarlem Oil, m. 1¼; Cinnamon Oil, m. ½,

In the treatment of that class of cases for which Methyloids are recommended, it has been found by every physician of experience that no *single* remedy fulfills the requirements of successful treatment. In Methyloids a happy combination has been formed, which will be found to possess peculiar merit. Methyloids are convenient to carry and to take, are readily soluble, the ingredients are the purest and are protected from deterioration, and the dosage is accurate.

Per doz.

Vials of 40. . . . .$4 50

BLUCALOIDS.
TRADE MARK.

Capsoids containing pure Australian Eucalyptus Oil, m. 3¾, and chemically pure Methylene Blue, gr. 1½.

Highly recommended in the treatment and prevention of malaria in its various manifestations, such as intermittent fever (chills and fever, fever and ague, malarial fever), remittent fever, bilious fever, malarial neuralgia, etc.

DOSE.—One Blucaloid four to eight times a day, or as directed by the physician.

Per doz.

Vials of 40. . . . .$4 50

APIOLOIDS.
TRADE MARK.

CAPSOIDS CONTAINING 5 M. PURE APIOL.

All authorities unite in ascribing to apiol valuable emmenagogue properties, and it has come into extensive use in the treatment of menstrual affections, whether there is an absence of the natural courses, or the period is attended with much pain.

Per doz.

Vials of 40. . . . .$6 40

DIGESTIVE FERMENTS
AND THEIR PREPARATIONS.

Pepsin-Stearns, 1-3000, U. S. P (In scales or powder)—1 oz. for 30 cts., ¼ ℔. for $1.00, ½ ℔. for $1.85, 1 ℔. for $3.50.

Saccharated Pepsin, U.S.P.—1 doz. ozs. for $1.50, ¼ ℔. for 30 cts., ½ ℔. for 55 cts., 1 ℔. for $1.00.

Glycerole Pepsin—1 pint for $1.00, 5 pints for $4.75, 1 gallon for $7.00.

Essence Pepsin-Stearns—½-pints, per doz., $4.00; pints per doz., $6.50; 5 pints for $2.40; 1 gallon for $3.60.

Lactinated Pepsin—1 doz. 1 oz. bottles for $3.00, ¼ ℔. for 75 cents, ½ ℔. for $1.45, 1 ℔. for $2.80.

Elixir Lactinated Pepsin—(38 grs. to oz.)—1 doz. pints for $5.50, 5 pints (bulk) for $2.00, 1 gallon (bulk) for $3.00. (80 grs. to oz.) per gallon $3.50.

Pancreatin, Pure—1 oz. for 40 cts., ¼ ℔. for $1.30, ½ ℔. for $2.50, 1 ℔. for $4.80.

Pancreatin, Saccharated—1 oz. for 24 cts., ¼ ℔. for 75 cts., ½ ℔. for $1.45, 1 ℔. for $2.80.

Pancreatin, Essence—½ pint bottles per doz. $5.00, 1 pint bottles per doz. $8.00, 5 pints (bulk) for $3.00, 1 gallon (bulk) for $4.50.

Rennin—1 oz. for 50 cts., 1 ℔. for $7.20.

Diastase-Stearns—¼ oz. vial for 45 cts., ½ oz. vial for 80 cts., 1 oz. vial for $1.50.

Essence Diastase-Stearns—½ pint bottles per doz., $6.00.

Panzyme—In vials of 50 tablets, per doz., $4.00.

NOTES OF REFERENCE

AND

TABLES OF INFORMATION.

DIGESTIVE FERMENTS

And Their Preparations

Peptic Starine, 1:3000, U. S. P. (Saccharated)...

Saccharated Pepsin, U.S.P. ...

Glycerin Pancreatin ...

Essence Pepsin Stearns ...

NOTES OF REFERENCE.

1. Acid Phosphates, Liquid. Each fluidounce contains: Calcium phosphate, 3 grs.; Magnesium phosphate, 1-2 gr.; Potassium phosphate, 1-4 gr.; Iron phosphate, 1-2 gr.; Syrupy Phosphoric acid, 6 3-8 minims. Dose, 1-2 to 1 teaspoonful in sweetened water. The stimulating properties of this preparation have long been known to physicians, and it has been extensively prescribed in various nervous disorders, including insomnia, melancholia, and nervous exhaustion. For price see page 186.

2. Aconitine, crystalline. In view of the fact that a number of preparations from aconite are sold as aconitine, and as there is a great variation in strength, we desire to call attention to the fact that we employ only the best grade of crystallized aconitine in our preparations. The crystalline alkaloids of aconite are extremely sensitive and are apt to become converted into the amorphous alkaloids which are in no way comparable to them as regards activity, being in fact considered inert by many authorities. The crystalline alkaloid when pure will be easily detected by the physiological test when dissolved in from 100,000 to 200,000 parts of water, and is therefore from two hundred to four hundred times as strong as the root.

3. Adhatoda Vasica, Malabar nut. An Indian plant which is alleged to possess a powerful bactericidal influence, but to be non-toxic to the higher forms of animal life. It is said to contain an alkaloid, vasicine, but has not been studied to any extent and consequently very little is known of its real value as a medicinal agent. For price of fluid extract see page 4.

4. Adonis vernalis. One of the Ranunculaceæ native to Northern Europe and Asia, which is being widely recognized as a cardiac tonic. It contains a glucoside, adonidin, to which its activity is usually ascribed. The drug resembles digitalis in action, but is considered more prompt and possesses the further advantage of having no cumulative effect. It is usually exhibited as fluid extract. For price see page 4.

5. Alterative Compound. The combination of vegetable alteratives, known under a variety of names such as Mc-Dade's Prescription, Bamboo-brier Compound, Creek Indian

Remedy, etc., and prepared from the formula of Drs. G. W. McDade and J. Marion Sims. It has been employed with considerable success in the treatment of primary and secondary syphilis. For price see page 5. We also direct attention to our Fl. Ext. Trifolium Compound (page 46), and Syr. Trifolium Compound (page 175), as especially valuable alterative combinations.

6. Antiseptic Tablets, Seiler's. Each tablet contains Sodium bicarbonate, Sodium borate, of each 3 3-4 grs.; Sodium benzoate, Sodium salicylate, of each 5-32 gr.; Eucalyptol, Thymol, of each 5-64 gr.; Menthol and Oil of Wintergreen, ot each about 1-24 gr.; and is intended to be dissolved in two ounces of water. A solution of this strength has been found very beneficial in nasal catarrh, being sufficiently alkaline to dissolve the thickened secretions adhering to the nasal mucous membrane, and, by its antiseptic properties, destroying the odor and bringing about a healthy condition of the passages. For price see page 139.

7. Arnica. We list fluid extracts of both the root and the flowers; the former is the official preparation, but as the latter is the one in most common use, we invariably supply the fluid extract of the flowers when fluid extract of arnica is ordered without further specification. See page 6.

8. Bamboo-brier. The rhizome of this plant which grows abundantly in southeastern United States is well known as an alterative and tonic, and is thought by many physicians to be superior in these respects to the sarsaparillas obtained from Central and South America. For price of fluid extract see page 7. See also note No. 5 on Alterative Compound, sometimes known as Bamboo-brier Compound.

9. Bay Laurel, Concentrated. We were the first to offer to the trade a concentrated fluid Bay Laurel for the extemporaneous preparation of Bay Rum. We have consistently maintained the quality of our product, and feel that we are justified in the claim that our concentrated Bay Laurel is unequalled by any other similar preparation now on the market. The addition of one and one-half fluidounces to one gallon of dilute alcohol produces a Bay Rum fully equal to the imported article, and at a much smaller cost, and, if this be again diluted one-half, the resuiting product will still be superior to that usually supplied to barbers. When prepared according to directions given on the label, the Bay Rum is ready to dispense immediately, no filtration being required. For price see page 7.

10. **Beef, Preparations of.** Most liquid preparations of beef, such as Beef, Iron and Wine, and its various combinations, are prepared from the ordinary commercial extract of beef. The popular impression that beef extract is highly nutritious and fully represents the beef from which it is made is wholly fallacious and is repudiated by the best authorities. Beef extract consists of the soluble salts and flavoring matters of beef and these are slightly stimulating but not nutritious. The nutritious portion of beef is its proteid matter, which is for the most part insoluble and consequently is almost entirely lacking in commercial beef extracts; therefore, while these latter may have an agreeable flavor, though even this is not often the case, their properties are only those of a condiment.

The preparations of beef which we offer differ from others of similar name because ours have nutritive properties and are *true foods*. Instead of using beef extract, which is worthless, we employ *beef peptone* which is made by peptonizing choice selected lean beef, thus obtaining its proteid, nutritious matter, including fibrin, in a soluble and acceptable form. Our preparations, therefore, have not only the agreeable flavor of beef but its nutritive properties as well. This point is well worthy the attention of both physicians and pharmacists who wish to supply a reliable and satisfactory preparation. For prices see page 177.

11. **Berberis Aquifolium.** This drug has become known as a valuable tonic and alterative and occupies a very important place in modern materia medica. Its principal active constituent is berberine, which is the yellow alkaloid occuring in hydrastis and numerous other drugs. Our fluid extract is assayed and warranted to contain not less than 3.25 per cent. berberine, and should therefore be given the preference over preparations which are not standardized, as the yield of alkaloid in Berberis Aquifolium often runs as low as one per cent. Obviously a fluid extract prepared from such a sample would have to be given in doses three times as large

as ours in order to produce the same effect, and would therefore be about three times as expensive, in addition to the disadvantage of its larger dose. For price see page 8.

12. Blackberry Aromatic. A reliable and satisfactory product for the extemporaneous preparation of Aromatic Syrup (or Elixir) of Blackberry, of unusual quality. The demand for a mild, pleasant and effective astringent and carminative for the treatment of summer complaints, such as colic, diarrhea and dysentery is well met by this combination. For price see page 9.

13. Black Cohosh Compound. This compound fluid affords a convenient means for the extemporaneous preparation of Compound Syrup (or Elixir) of Cimicifuga. The combination is a valuable one, having sedative, expectorant and antispasmodic properties. Although it is used chiefly as an expectorant, it has been found valuable in the treatment of uterine colic, locomotor ataxia, and in short, as a general antispasmodic. For price see page 9.

14. Black Willow. The bark of this tree contains about fifty per cent. more salicin and about twenty-five per cent. less tannin than that of the white willow. It is, therefore, more decidedly antiperiodic, and somewhat less astringent than white willow bark. Both bark and buds of black willow are considered useful in spermatorrhea and various forms of uterine and ovarian neuralgia, and recently they have come into favor as anaphrodisiacs. For price of fluid extract see page 10.

15. Blood=flower. Astringent, styptic, and anthelmintic. It has also been recommended in gonorrhea and vaginal leucorrhea. For price of fluid extract see page 10.

16. Boldo. An evergreen shrub native to Chile, whose leaves possess tonic and diuretic properties. It is considered of value in hepatitis and chronic torpor of the liver. In South America it has a considerable reputation in the treatment of gonorrhea and cystitis. For price of fluid extract see page 10.

17. Broom=corn seed. A demulcent diuretic and sedative, which exerts a peculiarly soothing influence over the genito-urinary tract, in inflamed conditions. It is entirely different from broom-top, Cytisus Scoparius, and should not be confounded with it. · For price of fluid extract see page 11.·

18. Buchu. Although both the long and short buchu are official and may be used without discrimination, it is the general belief that the latter is of greater therapeutic value than the other, and we accordingly supply fluid extract of short buchu unless the other variety is especially requested. See page 11.

19. Buchu, Compound Preparations of. We offer five different compound fluid extracts of buchu, intended for the extemporaneous preparation of compound elixirs, one of which is that of the N. F. These are combinations of well-known diuretics and antilithics, and will be found valuable in the treatment of diseases of the kidneys and bladder, inflammation of the genito-urinary tract, etc.

⸣ The preparations are: Buchu Comp. for Stearns' Elixir; Buchu Comp. for N. F. Elixir; Buchu, Juniper and Potassium acetate for Elixir; Buchu and Pareira Brava, for Elixir; and Buchu and Pareira Comp. for Elixir. For prices see page 11.

20. Caffeine citrate is a chemical possibility, but is difficult to prepare and is easily decomposed. In the presence of water it splits up into caffeine and citric acid, therefore it cannot be formed by mixing aqueous solutions of the acid and the alkaloid, as was formerly supposed. The Pharmacopœia recognizes this fact, and adopts the title Citrated Caffeine for a mixture of equal parts of caffeine and citric acid, which is to be dispensed whenever "caffeine citrate" is specified. Wherever the term Citrated Caffeine is used in this catalogue the U. S. P. product is the one referred to.

21. California Laurel. This is a California shrub whose leaves are said to be efficacious in nervous headaches, intestinal colic, and atonic diarrhea. The fluid extract is considered the best form for administration and the dose ranges from ten to thirty minims. For price see page 13.

22. Caroba. In Brazil, the habitat of this small tree, several species of the genus Jacaranda are known as caroba, carobo, etc. We employ Jacaranda procera in the preparation of our fluid extract. Caroba is extensively employed in Brazil as an alterative and antisyphilitic, and is also considered diuretic and diaphoretic. For price see page 14.

23. Cascara Amarga. This drug, native to Mexico and Honduras, is an alterative which is coming into favor in the treatment of many diseases of syphilitic origin, particularly syphilitic eruptions, gummatous tumors, chronic eczema, etc. For price of fluid extract see page 14.

24. Cascara Sagrada. According to Prof. J. U. Lloyd, in a paper read before the A. Ph. A. in 1896, the credit for introducing Cascara Sagrada to the medical profession is due to Dr. J. H. Bundy, of Calusa, Cal., in 1877. Although the drug was received with somewhat doubtful favor at first, its value has been so thoroughly demonstrated that it occupies a place in modern materia medica among the most important drugs. Inasmuch as it did not become very popular with the medical profession before a palatable fluid extract was placed on the market, and as we were the originators of the first and only fluid extract of it which is not bitter and does not gripe, we feel that much of its success as a remedial agent is due to our own efforts. We publish several pamphlets on this important drug which we shall be pleased to send to any physician or pharmacist on request. In addition to the preparations hereafter mentioned, we make the ordinary bitter fluid extract for those who for any reason may prefer it to the more palatable preparations. Our cascara bark is aged by storing it fully two years before using, by which means the griping property is considerably lessened. We wish to reiterate that cascara should be used only as a tonic laxative, and when in the treatment of chronic constipation cathartic effects are desired, some more active purgative should be employed, followed by small doses of cascara. See page 14.

Kasagra. This is a preparation of which we are justly proud. Its success has been almost phenomenal from its very introduction, and each year its sale has been largely increased. This could not have happened if it were not all we claim, for we have not expended much money in advertising it. For several years we listed this preparation under the name of Cascara Aromatic, but several other manufacturers, seeing the remarkable success it was achieving, adopted the same and similar titles for preparations of their own, which were not the same as ours, and none of which is of full fluid extract strength. In this way they did injustice to the medical profession and to us, and we do not believe that they have profited greatly by such a course. Therefore, in order that the medical profession, as well as ourselves, may be protected when they prescribe our preparation, we have adopted for it the name KASAGRA, which is a fanciful word coined by us to mean Cascara Aromatic–Stearns'. Prof. John Uri Lloyd, in a recent letter to us, said:

"To Frederick Stearns & Co., of Detroit, Mich. (1889), is due the credit of producing and bringing before the medical and pharmaceutical professions the first palatable fluid extract (Cascara Aromatic) of Cascara Sagrada. I consider this introduction so marked that out of credit to that house I refused to make a Cascara Aromatic, and when the house of Lloyd Bros. has orders for it, leaving the selection to us, the preparation of Frederick Stearns & Co. is invariably supplied, for to *us* the name of "Cascara Aromatic" means *their* preparation. When the history of the preparations of this drug is written, I believe full credit must be given to Frederick Stearns & Co. for that conspicuous discovery." For price see page 195.

25. Cascara Improved. The regular fluid extract of cascara is precipitated by the addition of water, and as it is frequently prescribed in combination with aqueous liquids an unsightly mixture is the result. To obviate this difficulty, which is one of pharmaceutical importance only, we have prepared Cascara Sagrada, Improved, which differs from the regular chiefly in being soluble in aqueous liquids. It is somewhat more agreeable to the taste than the regular fluid extract, but less so than Kasagra. For price see page 14.

26. Cedron. The seeds of Simaba Cedron, a tree growing in Central and South America, have a considerable reputation there as a remedy for the bites of snakes, poisonous insects, etc. Whether or not this reputation is deserved is as yet undecided, but the drug appears to possess valuable tonic, antiperiodic and antispasmodic properties. It is also a griping purgative if given in excessively large doses. For price of fluid extract see page 15.

27. Cereus grandiflorus. Inasmuch as clinical experience has shown that Cactus Bonplandii and Cactus McDonaldii have therapeutic properties identical with Cereus (or Cactus) grandiflorus, and as the medical profession seem to prefer it, we list only the latter named variety. Sharp and Hoseason did not succeed in finding either alkaloids or glucosides in this plant, although they examined a considerable number of specimens. They established the presence of several resins, however, all soluble in absolute alcohol, to which they are inclined to ascribe such activity as the drug may possess. For price of fluid extract see page 15.

28. Chapparro Amargoso. A reputed antiseptic, astringent and antiperiodic, much esteemed in diarrhea and dysentery. Its activity is supposed to be partly due to a resinous principle known as amargosin: The plant is of South American origin. For price of fluid extract see page 16.

29. Chekan. This evergreen shrub is indigenous to Chile: its leaves are considered efficacious in chronic respiratory catarrh, chronic bronchitis, and are also said to possess antiseptic, diuretic and expectorant properties. It is said to contain chekenon, chekenin and chekenetin, which are crystalline bodies ; a volatile oil and an amorphous bitter principle. For price of fluid extract see page 16.

30. Cinchona. Our fluid extract of cinchona is made to conform to the official requirements, namely, 5.% total alkaloids, of which not less than one-half is quinine. For red cinchona our standard is 5.% total alkaloids, and for pale cinchona, 3.% total alkaloids. We list also Detannated Cinchona, for use in mixtures containing iron salts; also Fluid Cinchona Compound for preparing U. S. P. Tincture, and Liquid Cinchona Compound for preparing B. P. Tincture. Both of these latter contain red cinchona. We offer also Fluid Cinchona Aromatic, which will be found a convenient combination for preparing aromatic wines, elixirs and tinctures of cinchona. See pages 16, 17.

31. Coca. The properties of coca leaves are by this time so well known as hardly to need any special mention. The drug has been shown to be of remarkable value in conserving energy and retarding tissue waste. It is a useful stimulant to the brain and nervous system in neurasthenia, melancholia, protracted mental depression, etc. Our fluid extract is made to conform to a standard of .5% cocaine, and is therefore uniform in strength. This is of importance, as there is a wide range of variation in the alkaloidal strength of coca leaves as found in the market, and a fluid extract which is not standardized is an uncertain factor in the treatment of disease. In addition to the regular fluid extract we offer a soluble fluid extract, for the preparation of elixir or wine of coca. For prices see page 18. See also "Cocavin," page 204.

32. Cocillana. A Bolivian drug which has been found to possess expectorant properties of considerable value, it being held by some to be fully as efficacious as ipecac in this respect. In large doses it acts as an emetic, producing considerable depression, dull headache, and sneezing. In acute and subacute bronchitis, and bronchial pneumonia, it has been employed with some success. The active principle is thought to be a glucoside, but has not been thoroughly investigated. For price of fluid extract see page 18.

34. Corn Ergot. This drug has given fair promise of rivaling ergot as an oxytocic, but as yet its use has not been very extensive. Its contracting action on the uterus is said to be intermittent, differing in that respect from the ordinary (or rye) ergot. It is said also to be useful in eczema and psoriasis. For price of fluid extract see page 19.

35. Corn-silk. Though this drug is now officially known as Zea we list it as corn-silk, that being the name by which it is almost universally called. It is a mild, stimulant diuretic, which has been found to be quite valuable in acute and chronic cystitis, and in irritation of the bladder due to uric acid and phosphatic gravel. It has also been employed in the treatment of gonorrhea. For price of fluid extract see page 19.

36. Coto. The botanical origin of this Bolivian drug is as yet undetermined. Our fluid extract is made from true coto, instead of false or para-coto, and as para-coto probably is less active, this is an item which should be remembered in prescribing and dispensing. Coto is an intestinal antiseptic and astringent and is highly recommended in diarrheas of various forms particularly those of phthisis, typhoid fever and cholera. It is also said to be useful in rheumatism. For price of fluid extract see page 19.

NOTES.

37. Damiana. This drug, which was introduced as a powerful aphrodisiac, has not fulfilled the expectations of the medical profession, but has nevertheless been found to be of some value in atonic impotence. It probably acts principally as a bitter tonic. For price of fluid extract see page 20.

38. Damiana Compound. This preparation, the formula for which is given in the fluid extract list (page 20), will be found a valuable tonic to the sexual organs. It is a cerebrospinal stimulant and is extensively used in the treatment of impotence due to atony.

39. Diastase-Stearns. Diastase-Stearns is the name we have given to our brand of pancreatic diastase to distinguish it from other brands of diastase on the market. We take great pains in its preparation, having established a laboratory in Chicago, where we obtain a large supply of fresh pancreas. Being at this great central market gives us opportunities that cannot be obtained by those who are obliged to ship their crude material for long distances. Our laboratory facilities are all modern, and, we believe, turn out a finer product than any other manufacturing establishment in the same line. That this is not a mere assertion can be easily demonstrated by comparing the physical and chemical properties of Diastase-Stearns with other products of a similar nature on the market. Ours is perfectly sweet in smell and taste, and possesses a starch-digesting power above that of any other diastase on the market, animal or vegetable.

In this connection it is of interest to note the statement made in a recent paper of Dr. H. W. Wiley (Chief of the Division of Chemistry, U. S. Dept. of Agriculture,) and W. H. Krug, read before the American Chemical Society, and published in the April Journal of that organization. The paper treats of the "Estimation of Starch," and says concerning our product: "We have also tried quite extensively the pancreatic diastase prepared by Frederick Stearns & Co., Detroit, and found it exceedingly active and free of any reducing matter. The indications are that it may prove to be the most desirable form of hydrolyser yet used."

There are other sources of diastase beside the pancreas of animals. Diastase takes an important part in fitting the starch of vegetables for vegetable food. That is, vegetables store up starch in their tissues as food for future use, and before the plant can utilize it, conversion into sugar by the action of vegetable diastase must occur. But the needs of plants in this connection are very different from the needs of animals. Plants have plenty of time to digest their food, but one meal rapidly succeeds another in animals, especially in man. On this account nature has furnished the animal body with a diastase which is many times more active than vegetable diastase. Animal diastase is, therefore, better fitted for the digestive processes in the animal body. Again, as pointed out by Chittenden, the products of digestion of starch constitute a peculiar series, and each one of the series may have an important part to play in the nutrition of the body. As the action of diastasic ferments still remains, to a great extent, an unexplored field, it is impossible to say just what part these different products play in nutrition. There are, therefore, abundant reasons why animal diastase is to be preferred to that derived from plants. The latter is fitted to take a part in raising mineral com pounds to the plane of vegetable life. The former takes part in raising vegetable matter to that higher and infinitely more complex plane of existence peculiar to animals.

For price of Diastase-Stearns see pages 202, 203.

40. Digitalis. There are few drugs of greater importance to the medical profession than digitalis, which is without doubt the most extensively used cardiac tonic, and is perhaps the most reliable. It is therefore very necessary that the preparations of it should be active and uniform in strength. Our fluid extract of digitalis is not excelled in these two important requirements by any other preparation now on the market. It is made from prime, selected digitalis, and carefully adjusted by assay to a uniform standard of strength, so that each lot shall represent not only the full activity of the drug, but shall be of the same strength as every other lot of fluid extract of digitalis sent out by us. We commend it to the medical profession with the confidence which comes from a knowledge of its high therapeutic value and the satisfaction with which it has ever been used in the past. For price see page 20.

41. Dover's Powder, Fluid Extract. This preparation, which is often called Fluid Dover's Powder, is simply the official Tincture of Ipecac and Opium. Each Cc. represents 1-10 gramme each of ipecac and opium, both of which are assayed before being used to make this preparation. The tincture may be used wherever Dover's Powder is indicated. For price see page 27.

42. Duboisia. The alkaloid of this drug, which has been called duboisine, has been found to be a mixture of hyoscine and atropine. The properties of duboisia are very similar to those of hyoscyamus or stramonium, although it is more of a cerebral sedative than either. It is said to be of considerable value in vesical tenesmus, night-sweats of phthisis, etc. For price of fluid extract see page 21.

43. Echinacea. The root of Echinacea angustifolia, more commonly known in the western United States as "Black Sampson," has been recommended as a very powerful alterative and antisyphilitic. It is also alleged to be useful in the treatment of wounds inflicted by snakes and rabid animals. For price of fluid extract see page 21.

44. Embelia. An East Indian drug which is reputed to possess valuable tenicidal properties, and to be almost specific in rheumatic complaints. According to Worden the active principle is embelic acid, and the ammonium salt of this acid was found an effective tenifuge. The drug has not been sufficiently investigated as yet for a correct estimate of its value to be reached. For price of fluid extract see page 21.

45. Ephedra antisyphilitica. An herb indigenous to western United States, particularly Arizona and contiguous territory. A somewhat superficial examination made by Loew did not reveal the presence of an alkaloid, but a peculiar tannin was separated. The drug enjoys a popular reputation in Arizona as a specific in venereal diseases, particularly gonorrhea, for which it is employed in the form of a decoction. It is probably only alterative and astringent. For price of fluid extract see page 21.

46. Ergot. Few drugs are more extensively employed than ergot, and it is therefore of great importance that preparations of it shall be of full strength. Our fluid extract is made from carefully selected latest crop ergot, and is carefully adjusted by assay, so that uniformity of strength as well as a high degree of therapeutic activity is assured. These qualities have won an enviable reputation for our fluid extract of ergot in the past, and we shall continue to keep it up to its present high standard of excellence. For price see page 21.

47. Eschscholtzia Californica. This plant, native to California and known as " California Poppy," is alleged to be a harmless calmative, analgesic and soporific. In addition, it is said to be free from certain effects which accompany the use or opium, such as constipation and derangement of digestion. It is one of the Papaveraceæ, and was claimed by Bardet and Adrian to contain morphine. This, however, failed of confirmation by König and Tietz, who could not find morphine in specimens of the plant cultivated at Marburg. For price of fluid extract see page 21.

48. Euphorbia. A tropical plant, said by Dujardin-Beaumetz and other clinicians to give excellent results in the treatment of asthma and asthmatic bronchitis, also in chronic and subacute bronchitis, as well as other affections of the respiration. For price of fluid extract see page 21.

49. Fluid Extracts, Assayed. Properly speaking, an assayed fluid extract is one which has been submitted to appropriate tests and made to conform to a certain standard. In the cases of those drugs for which the pharmacopœia specifies standards the official requirements are taken as the criterion, but the number so treated by the pharmacopœia is insignificient when compared with the total number—more than five hundred—listed by us. Therefore we have adopted arbitrary standards in the other cases and in this we have been guided by our experience of many years in this line, so that our requirements for these are such as to make the finished product in every case represent not an average but a *prime* drug of the market.

It is true that comparatively a small proportion of the drugs known to modern pharmaceutical science may be assayed for alkaloids, but in the cases of non-alkaloidal drugs we assay for glucosides, resins, oleoresins, etc., depending on the nature of the drug, i. e., what its active principle may be. In still other cases, and these of course are numerous, we are compelled to resort to a standard for dry extractive, inasmuch as scientific investigation has not yet disclosed more accurate means of valuation. Some few drugs also may be more accurately standardized by physiological tests; conspicuous among these is aconite.

Our methods of assay are of course varied to meet the requirements of individual . drugs. ' For alkaloids we employ principally volumetric methods of estimation, but in some cases gravimetric; of the volumetric methods we resort to titration with standard acids in some cases, while in others we find titration by Mayer's reagent to be more satisfactory.

It will be seen then, that our fluid extracts (and the same applies to our extracts) may be relied on to represent the drugs they are made from, truly and exactly. There is no haphazard, no guesswork about our methods: beginning with an assayed drug, assaying the unfinished product and finally assaying the finished preparation we obtain a fluid extract which is a model of pharmaceutical skill.

50. Fluids. Among the numerous other new pharmaceutical preparations that we have from time to time introduced are the fluids (listed among, but not necessarily fluid extracts,) which represent concentrated forms of various official and unofficial preparations, such as tinctures, syrups, elixirs and wines. These are carefully prepared, and on being diluted according to directions on the label furnish elegant and permanent preparations equal in all respects to those obtained by following the official processes. We have sometimes called these preparations " Druggists' Conveniences," and we feel sure that the thousands who have used them will agree with us that they are conveniences—and necessities as well. Since our successful introduction of this line of preparations, other houses have followed our example and offer similar goods to the trade. We do not believe, however, that any other house in the United States offers so complete a line of them. Nearly every compound tincture and compound syrup of the U. S. and British Pharmacopœias, most of the important elixirs, tinctures and syrups of the Eclectic Dispensatory and National Formulary, and many other unofficial but widely used preparations may be prepared from these fluids by simply diluting with the proper menstruum, be that alcohol, elixir, syrup or wine.

In addition to this line of preparations, several other compound fluids are offered which are intended for . medicinal use without further manipulation. In most cases they are not true fluid extracts, but for the sake of convenience are listed along with them.

51. Fluids, Half-Strength. We list quite a number of preparations among our fluid extracts for the sake of convenience, which are not true fluid extracts. Our fluids to which reference is made under note No. 50 are examples of this, also some tinctures which are given among the fluid extracts, and in addition thereto a class of preparations known generally as half-strength fluids. As is well known to most pharmacists, it is impossible to make permanent, true fluid extracts of certain drugs consisting largely or wholly of resins, gum-resins, etc. But there is a demand for concentrated liquid preparations of such drugs and this demand we supply with our "half-strength" or "semi-normal" fluids. In most cases these are just one-half the strength of true fluid extracts, i. e. each Cc. of the fluid represents one-half gramme of the drug. Among these may be mentioned aloes, asafetida, benzoin, catechu, myrrh, and tolu. Besides these we prepare a Fluid Guaiac resin, of 60 per cent. strength, and a Concentrated Opium for tincture, of 40 per cent. strength. These fluids are very convenient for the druggist's use in making up tinctures, etc., and will be found to answer the purpose completely.

52. Gelsemium. Several years ago we dropped from our list the fluid extract of the green root, as clinical investigations had failed to demonstrate that there was any advantage possessed by it over that of the dried drug. In fact, most of the testimony is to the effect that the fluid extract of the green drug is less reliable than the other. Our fluid extract of gelsemium is assayed carefully several times during its process of manufacture and each lot is made to conform to the same standard, viz., .5 per cent. gelsemine. This insures uniformity of strength and a high standard of activity. For price see page 23.

54. Golden Seal, Colorless. This preparation, which has been on the market for a number of years, owes most of its medicinal value to the white alkaloid of hydrastis which it contains. It is largely used as an injection or wash in inflamed conditions of the mucous membranes, and being colorless does not stain the clothing. For price see page 24.

55. Golden Seal, Non-alcoholic. To meet the demand for a fluid aqueous preparation of golden-seal we placed this on the market several years ago. It contains both the white and yellow alkaloids of golden-seal together with the other principles of the root. It is miscible with both aqueous and alcoholic liquids without precipitation and is an elegant and efficient preparation of the drug. For price see page 24.

56. Guaco. In Mexico, and Central and South America, guaco is considerably employed as a febrifuge and anthelmintic, and is also alleged to be of considerable value in rheumatic affections, epidemic cholera and chronic diarrhea. For price of fluid extract see page 24.

57. Guarana. We have adopted a standard of 4.% caffeine for the fluid extract of this drug, although that is somewhat above the average yield of caffeine in the guarana commonly found in the market. On account of its comparatively large alkaloidal content, this drug has come into considerable prominence as a remedy for sick headache and for other nervous complaints in which caffeine is indicated. For price see page 24.

58. Helonias Compound. Although other manufacturers use the term Helonias Compound as a synonym for Mitchella Compound (Squaw Vine Compound) of the Eclectic Dispensatory, we, as introducers of the *original* Helonias Compound, retain the original name for our formula, in which helonias predominates. We list also Mitchella Compound, of the Eclectic Dispensatory, under its proper title, but desire to call special attention to the fact that the medical properties of Helonias Compound and Mitchella Compound are quite different, the latter being a parturient, emmenagogue and antispasmodic, while Helonias Compound is a genito-urinary tonic and diuretic. For prices see pages 25, 164.

59. Horse-nettle. A perennial plant indigenous to southern United States, which has come into prominence on account of its value in epilepsy. It is also useful in idiopathic tetanus and in convulsions accompanied by albuminuria during pregnancy. Aphrodisiac properties have also been

ascribed to it. We offer fluid extracts of the root and berries.
For prices see page 26.

60. Hydriodic Acid, Concentrated Solution. This
solution is ten times as strong as the official syrup, therefore
the latter may be prepared from it by adding nine parts (by
weight) of simple syrup to one part of the solution. Occasion-
ally, also, physicians wish to prescribe two or three per cent.
syrup of hydriodic acid. With the concentrated solution, a
syrup of the desired strength can be prepared by altering the
proportions used. For price see page 156.

61. Hydriodic Acid, Syrup of. Our syrup of hydriodic acid conforms completely to U. S. P. requirements, and
is not surpassed in permanency by any other preparation of
its kind on the market. For price see page 172.

62. Hysterionica. An herb native to Chile, which is
said to be of considerable value in flatulent dyspepsia and
in inflammation of the bowels accompanied by hemorrhage.
Baille considers it also a reliable remedy in gonorrhea and
other complaints involving inflammation of the genito-urinary
tract. For price of fluid extract see page 26.

63. Indian Hemp. Four or five different drugs are more
or less generally known as Indian Hemp, either with or without
other descriptive titles. This often leads to confusion and
might possibly cause serious results by the administration of
one drug under that name when another was desired. We
have, therefore, decided to restrict our use of the term Indian
Hemp to the official Cannabis Indica from Cannabis sativa.
Apocynum, sometimes called Black Indian Hemp, is listed by
us under the name Canadian Hemp. Asclepias incarnata, sometimes called White Indian Hemp, is listed as Swamp Milkweed. Therefore, when we receive orders for fluid extract of
Indian Hemp, we shall supply the U. S. P. preparation unless
otherwise requested.

Indian Hemp, Powdered Extract. We have satisfied
ourselves by experimentation that powdered extract of Indian
Hemp cannot be so prepared as to represent more than *one-
third* of an equivalent weight of the pilular extract. Our
preparation, therefore, conforms to that standard, being an ex-

ception to the general rule that our solid and powdered extracts are of the same medicinal strength, weight for weight. For price see page 59.

64. Jamaica Dogwood. This drug which we introduced to the medical profession, has proven a valuable anodyne, sedative and hypnotic. The bark of the root is the part employed, it being much more active than other portions of the tree. For price of fluid extract see page 27. •

65. Jambul seed. The remarkable property of arresting the conversion of starch into sugar is attributed to this drug, upon somewhat slender evidence it would seem, and it is said to have been used with success in diabetes, serving to diminish considerably the amount of sugar excreted. In India it is also used as a stimulant and astringent in diarrhea. For price of fluid extract see page 27.

66. Jurubeba. This Brazilian drug is alleged to be of value in gonorrhea and in syphilis, but its principal use is as a deobstruent and tonic. For price of fluid extract see page 28.

67· Kamala. The value of this most efficient tenicide is well known. It acts particularly well in combination with male-fern, and we wish to call attention to our No. 49 soft elastic capsules of male-fern and kamala, page 154, and No. 71 extra sized soft elastic capsules of male-fern, kamala and castor oil, page 155. For price of fluid extract see page 28.

Kasagra. See note 24.

68. Kava-Kava. Has been employed with good results in the treatment of gonorrhea, vaginitis, leucorrhea, and in other inflamed conditions of the genito-urinary tract. For price of fluid extract see page 28.

69. Kino, Tincture. We formerly listed a half-strength fluid kino, for the extemporaneous preparation of the tincture, but discontinued it as we found it liable to gelatinize, causing our customers and ourselves considerable annoyance. We now offer the official tincture of kino, which is much more permanent and satisfactory than the more concentrated fluid preparations of the drug. For price see page 28.

70. Kola. We are the introducers of kola in America, and the chemistry of this drug which has become so important in modern materia medica, has been investigated to a greater extent through our instrumentality than that of any other house in the United States, we might almost say, in the world. We were the original, and are yet the largest importers cf the fresh, undried, true African kola into this country. We have published more scientific literature on kola than any other house and will be pleased to send to anyone who may desire it, our illustrated monograph, also reprints of scientific papers describing the recent investigations into the chemistry of this interesting drug. Kola has fully justified the claims made for it as a cerebro-spinal nerve stimulant and con-servator of energy. We are in possession of a great num-ber of clinical reports certifying to the value of our prepa-rations of kola in nervous exhaustion, lassitude, melancholia, mental depression, cardiac weakness, vomiting in preg-nancy, etc., etc., indicating that its scope as a remedial agent is wide and that it is a trustworthy remedy when em-ployed in the form of preparations of the fresh. nut, such as Kola-Stearns, which is a palatable, full strength fluid extract, or Kolavin, which is a pleasantly aromatized wine of fresh kola. We also offer Kolacyls, a delicious confection of the active medicinal principles of kola as they exist in natural combination undisturbed by the addition of other chemicals; these are designed for the use of bicyclists and athletes, being put up in very con-centrated form, convenient to carry in the pocket.

Finally let it be said once for all that there is not the slightest danger of a "kola habit" being formed. Although the drug has been used very extensively in the last few years, almost indis-criminately by the laity in some cases, there is not as yet one case on record tending to show that a "kola habit" has been formed. Furthermore there is no reason whatever to suppose that such a habit would be any more deleterious than that of tea or coffee which is admittedly not of a very serious nature. See pages 28, 200, 201.

71. Kousso. This drug is highly valued in Abyssinia as a vermifuge, and has been proven to have extraordinary efficacy in the destruction of tapeworm. For price of fluid extract see page 28.

71 a. Lactinated Pepsin. Few combinations of digestive ferments have given better satisfaction than this one. It contains pepsin, pancreatin, diastase, lactic acid, hydrochloric acid and milk sugar, thus representing the various digestive fluids of the body. The liquid preparations include Liquid Lactinated Pepsin, Syrup of Lactinated Pepsin with Phosphates and a number of Elixirs, all of which possess the therapeutic properties of the various amounts of Lactinated Pepsin contained in them. For prices see pages 179, 180.

72. Licorice Aromatic. We offer three aromatic preparations of licorice, viz: the fluid extract, syrup and elixir. The fluid extract (see page 29) is itself an excellent vehicle for the administration of quinine and other bitter substances, and from it, by diluting properly, an aromatic syrup or elixir can be prepared. The syrup (see page 173) and the elixir (see page 167) offered in our list are elegant preparations capable of masking 25 grains of quinine to the fluidounce.

As the sweet principle of licorice is precipitated by acids care should be taken not to dispense acids or acid salts in combination with either of these preparations.

73. Lily of the Valley. In addition to the official fluid extract of the root we list also fluid extracts of the herb and flowers. This drug, whose action as a cardiac tonic is very similar to that of digitalis, is coming into extensive use and has the advantage over digitalis of having no cumulative action. All parts of the plant possess the same medical properties but the root seems to contain the largest amount of active principles. For prices of fluid extracts see pages 29, 30.

74. Lozenges, Compressed, Weight of. Following is given a list of those compressed lozenges which are not of the usual size (20 grains each, 350 to the pound). For price list of lozenges and troches see page 144.

List No.		Weight in grains.	Number per ℔.
3.	Ammonia, Jackson's	10	700
4.	Ammonium chloride, "A"	12½	560
5.	Ammonium chloride, "B"	10	700
16.	Catechu, U. S. P.	11¼	622
17.	Chalk, U. S. P.	11	636
19.	Charcoal, 10 grs.	11	636
20.	Charcoal, 20 grs.	22	318
27.	Cubeb, U. S. P.	7	1,000
32.	Ginger, U. S. P., "A"	22½	311
36.	Ipecac, U. S. P.	10½	667
38.	Iron, reduced, B. P.	17	412

List No.		Weight in grains.	Number per ℔.
39.	Licorice, Ext., 5 grs.	5	1,400
40.	Licorice, Ext., 10 grs.	10	700
42.	Licorice and Opium, U. S. P.	7	1,000
43.	Magnesia.	12	584
44.	Morphine and Ipecac, U. S. P.	10	700
63.	Potassium chlorate, 5 grs.	25	280
64.	Pot. chlorate, 5 grs. (vanilla)	25	280
66.	Quinine tannate, 1 gr., "A"	10	700
67.	Quinine tannate, 2½ grs., "B"	10	700
68.	Rhatany, Ext., U. S. P.	11¼	622
70.	Rhubarb Blocks	50	140
80.	Santonin, 1 gr.	25	280
81.	Santonin, 1 gr., with chocolate	25	280
90.	Sodium bicarbonate, U. S. P.	12¼	571
91.	Sodium santoninate, 1 gr.	21½	326
96.	Tannic acid, B. P., ½ gr.	16	438
97.	Tannic acid, U. S. P., 1 gr.	11½	609
104.	Wistar's	10	700

75. Male-fern. There are but few drugs which deteriorate more rapidly than this one in the dried state and therefore many of the preparations of male-fern now on the market are quite inert and worthless. In preparing our fluid extract, the crude drug is first assayed and then the finished fluid extract is made to conform to a standard of 10 per cent. oleoresin. An active and thoroughly reliable product is thus obtained whose results will be found uniform. For price see page 31.

76. Manaca. This is a Brazilian plant the root of which is much employed in that country in the treatment of chronic subacute rheumatism and erysipelas. It has also been employed in this country to some extent, particularly in the treatment of rheumatic affections, with fairly satisfactory results. On account of its powerful alterative properties it has been sometimes called "vegetable mercury," but this title is more commonly understood to belong to mandrake, and to prevent confusion its use should be restricted to the latter. For price of fluid extract see page 31.

77. Muirapuama. A Brazilian drug reputed to be one of the most powerful aphrodisiacs known. Very little is known of its real value, as it has not yet received a thorough clinical investigation. For price of fluid extract see page 32.

78. Mutisia viciaefolia. A Bolivian drug belonging to the Compositæ which is said to be a valuable antispasmodic

useful in hysteria, also in croup and other affections of the respiratory apparatus. For price of fluid extract see page 32.

79. Newbouldia. The root bark of Newbouldia lævis,
native to tropical Africa, and said to be a valuable astringent
in diarrhea and dysentery. For price of fluid extract see
page 33.

80. Ointments, Ground. There are a number of ointments in general use which are extremely difficult to prepare in
the usual way. Conspicuous among these are the ointments of
zinc oxide, red and yellow mercuric oxides, and ammoniated
mercury. Our ointments are ground with machinery especially
constructed for the purpose, and are perfectly smooth and
homogeneous. They therefore give perfect satisfaction and
are elegant pharmaceutical preparations.

Inasmuch as we do not employ the ointment bases prescribed
by the pharmacopœia, the designation "U. S. P." signifies only
that the official standard of strength is adhered to. For list of
ointments see page 191.

81. Oleates. Most oleates are true chemical combinations,
and not merely solutions of the alkaloid or metal in oleic acid.
Oleate of arsenic, is however, an exception to this rule, it now
being considered that the oleic acid acts only as a solvent and
that the arsenic exists in the "oleate" as arsenous acid.

Our oleate of mercury is permanent and will keep in any
climate. This is an item of considerable importance, as
mercuric oleate is usually unstable.

It has been thoroughly demonstrated that medicinal agents
exhibited as oleates, or as oleic acid solutions, are much more
readily absorbed by the skin than when administered in combination with the ordinary ointment bases. For list of oleates
see page 189.

82. Opium, Preparations of. In addition to the official
wine and tinctures of opium, we offer Fluid Opium, Concentrated, assayed, of two-fifths regular fluid extract strength, for
the extemporaneous preparation of U. S. P. and B. P. tinctures; Fluid Opium, aqueous and deodorized, of the same
strength, for preparing Tinct. Deodorized Opium, U. S. P.;
Fluid Opium, camphorated (eight times the strength of
"Paregoric"), for preparing Camphorated Tincture of Opium,

U. S. P.; and Liquid Camphor Compound (eight times as strong as Comp. Tinct. Camphor, B. P., which it is used to prepare). For prices see pages 33, 34.

All our preparations of opium are assayed to conform to the official requirements, and are strictly reliable.

83. **Opium, Crystal.** This is an elegant scale or "crystal" preparation of opium (freed from narcotine and the disagreeable odorous principle), and is intended for dispensing purposes. It is in fine light-brown scales, and is assayed to conform to a standard of 10 per cent. morphine, 0.25 per cent. codeine, and 0.10 per cent. narceine. Prices quoted on application.

84. **Paraguay Tea.** A caffeine-bearing drug used in the interior of South America as a beverage in much the same way as tea and coffee are usually employed. It possesses stimulant and diuretic properties, and is also mildly astringent. For price of fluid extract see page 34.

85. **Passiflora incarnata.** Passion flower. This is alleged to be a valuable narcotic, useful in neuralgia, insomnia, dysmenorrhea, also in diarrhea and dysentery. For price of fluid extract see page 35.

86. **Pepsin, U. S. P.** Our pepsin we believe to be unsurpassed by any other now on the market. It is in beautiful translucent scales, which are free from odor, are readily soluble and non-hygroscopic. Being absolutely free from mucus, it yields a beautifully clear solution, and will be found far superior to the ordinary varieties for the preparation of elixirs, essences, wines, etc. It is guaranteed strictly 1-3000, and in every way fulfills the requirements of the U. S. Pharmacopœia. For price see page 208.

87. **Pichi.** A terebinthinate diuretic, reputed also to possess cholagogic properties. It is employed with benefit in both acute and vesical catarrh and is recommended in the treatment of gonorrhea and gonorrheal prostatitis. It is said to aid in the expulsion of renal and cystic calculi, but is generally considered to be contra-indicated in organic kidney diseases. For price of fluid extract see page 35.

88. **Pills—Friability and Solubility.** That the hardness of a substance has nothing to do with its solubility is very evi-

dent, on reflection. Taking some familiar examples, such as rock salt, rock candy (crystallized sugar), fused calcium chloride, potassium hydrate, sodium hydrate, and gelatin, we find that all are very hard and may be easily driven into a pine board without fracture. And these are among the most soluble substances known, consequently their hardness does not affect their solubility, or they would be quite insoluble.

On the other hand, the mere fact that a substance can be easily reduced to a powdered form does not in any way indicate that it is soluble. Some of the most insoluble substances known, such as charcoal, silver chloride, calomel, magnesia, prepared chalk, etc., etc., are very friable. We thus have the most convincing proof that soluble substances may be very hard, and that insoluble substances may be very friable. Therefore the assertion that mass pills are insoluble because hard, and that other kinds of pills are soluble because friable, is not in accordance with the facts.

Mass pills are usually prepared by massing the medicinal constituents with an excipient. If the pills are properly made this excipient is very soluble, and is composed chiefly of sugar, gelatin, acacia, dextrin, etc., in varying combinations and proportions to suit the individual requirements of each case. The coating also is either of sugar or gelatin. Therefore all the non-medicinal ingredients are soluble, and are, furthermore, *substances whose solubility is not in the least diminished by age*, as every pharmacist well knows. There is consequently not a shadow of foundation in fact for the assertion (made for commercial purposes) that all mass pills deteriorate with age. Some mass pills may—but ours, being properly made, will not and we are prepared to prove it. Our pills begin to disintegrate soon after reaching the stomach or duodenum (depending on the kind of coating, etc.), and produce their medicinal effect. This is because they are mass pills, soluble and properly made.

Friable pills, which are coated with a thin, brittle, insoluble shell, must be pulverized before they can be digested by the stomach. The absurdity of such a scheme is very apparent, for while these pills may be crushed to a powder under the thumb, it must be remembered that the digestive apparatus works, not by grinding like a drug mill, but by gentle agitation in the presence of solvent fluids. Therefore, insoluble substances which require crushing pass out of the alimentary canal with the waste products of the system. There is no more familiar example of this than the peanut, which if swallowed whole would pass through the alimentary canal intact. And yet a peanut is friable and may be easily crushed to fragments under the thumb. What is true of the peanut is true of friable pills, and if friable pills are to be administered they should be pulverized before swallowing, if any medicinal effect is to be expected.

Our contention is that mass pills, such as made in our laboratory, represent the highest development of the pill-makers' art. They are soluble, elegant, permanent, and reliable. We will be pleased to send copies of the Pharmacologist or New Idea containing Prof. Sherrard's article entitled "Friability is not Solubility," and describing the result of over 3000 tests of the solubility of the various pills of the market, to any physician or pharmacist, on application.

89. Pills, Gelatin-coated only. There are a number of pills whose bulk would be so materially increased by sugar-coating, that they would be inconveniently large and difficult to swallow, while gelatin-coating does not add appreciably to their size. For that reason we offer such pills in gelatin coating only.

90. Pills, Sugar-coated only. Gelatin coating is not suitable for such pills as undergo variations in consistency or volume by reason of changes in temperature, because they are liable to lose their shape or burst the coating in warm weather. Pills containing salts, either deliquescent or freely soluble in aqueous fluids, or ingredients which act unfavorably on the gelatin, rendering the coating insoluble, should not be gelatin-coated. For these reasons we offer pills belonging to the above named classes with sugar coating only.

91. Podophyllin, soluble. A soluble form of podophyllin, which, being nearly tasteless, is very satisfactory for dispensing in liquid preparations. It is incompatible with acids, metallic salts, and strongly alcoholic preparations. For price see page 176.

·92· Powdered Extracts, Assayed. Our powdered extracts are of equivalent medicinal strength with solid extracts, except that of Cannabis Indica, which is only one-third that strength as mentioned in note 63.

Powdered extracts of alkaloidal drugs are submitted to assay and made to conform to the standards specified in the list. The importance of this will be readily appreciated on considering the fact that the yield of extractive matter varies widely with different samples of the same drug. See page 59.

93. Quebracho. This drug, although said to be employed in Chile as an antiperiodic, is used in this country in the treatment of asthmatic affections. It stimulates the respiratory centers, and has been found by various clinicians to be a valuable remedy in emphysema, bronchitis, chronic pneumonia, etc. For price of fluid extract see page 37.

93 a. Rhubarb Blocks. Select rhubarb compressed into "blocks" or pieces of convenient pocket size. Put up in glass-top boxes holding one pound each. For price see page 148.

94. Rhus aromatica. Highly recommended for nocturnal incontinence of urine, hematuria and uterine hemorrhage. Has also been used with success in the treatment of diarrhea and dysentery.

95. Saw Palmetto berries. Although this drug has not long been before the medical profession, it has come into great prominence. It is a diuretic, tonic and genito-urinary vitalizer, and has a special action on the ovaries, testes, prostate, and mammary glands. Our fluid extract is prepared from fresh ripe berries gathered under the direct supervision of our own collectors. We will be pleased to send descriptive literature on request. For price see page 41. See also Triti-palm, page 197.

96. Senna. Although both the Alexandria and India sennas are official, we find that the preference of both physicians and pharmacists is for the Alexandria, and unless otherwise specified on the order, the fluid extract we supply is prepared from this variety. For price see page 41.

97. Senna pods. These have the same properties as the leaves, but are said to be slightly more active and less apt to gripe. For price of fluid extract see page 42.

98. **Sierra Salvia**, Mountain Sage. This drug was introduced as a substitute for quinine and has a local·reputation in its habitat, western United States, as an active and efficient diuretic, diaphoretic and antiperiodic. For price of fluid extract see page 42.

Solid Extracts. See note 107.

99. **Strophanthus.** Professor Fraser introduced this drug to the medical profession in 1870, and it is considered a most valuable cardiac tonic, acting similarly to digitalis, but being non-cumulative. It is exceedingly useful in the treatment of Bright's disease, for the dyspnea, orthopnea, dropsy and uremia. It is also said to have been used with good results in exophthalmic goitre. We list the official tincture, of which the dose is 3 to 10 minims. For price see page 44.

100. **Tonic Hypophosphites.** An elegant syrup of the principal tonic hypophosphites, properly prepared, and therefore free from any precipitate. Each fluidounce contains:

Calcium hypophosphite	. . . .	1 gr. ·
Potassium "	. . . .	1 1-2 grs.
Iron "	. . . .	1 1-2 grs.
Manganese "	. . . .	1 gr.
Quinine "	. . . .	1-4 gr. ·
Strychnine "	. . . .	8-100 gr.

Valuable in diseases of the pulmonary organs, such as bronchitis, asthma, coughs, colds and consumption; as a nutritive stimulant to the nervous system, stimulating the appetite and toning up the system. For price see page 172.

101. **Trifolium Compound.** This alterative compound, devised by Professor Rush, of Chicago, in 1885, is largely used as a tonic and alterative in a variety of skin diseases of venereal origin, also in syphilis of all stages, scrofula, eczema, ·chronic ulcers, glandular affections, chronic rheumatism, etc. It may be administered either as fluid extract or as syrup; the latter may be prepared by mixing a given amount of the fluid extract with three times its volume of simple syrup. For price of fluid extract see page 46: for price of syrup see page 175.

103. **Viburnum Compound.** This compound fluid extract, intended to be used in the preparation of the N. F. Compound Elixir of Viburnum Opulus, may be administered either as a fluid extract or as an elixir. It is a highly satisfactory antispasmodic, diuretic and tonic, and exerts a sedative

influence on the uterus. It is specially useful in nervous diseases and manifestations of pregnancy, threatened and habitual abortion, dysmenorrhea, amenorrhea due to nervousness and cold, after-pains and neurotic conditions of the menopause. It has also been used in hysteria. For price of fluid extract see page 47.

104. White Pine Compound. No combination of expectorants has given better satisfaction than this one in the treatment of chronic pectoral complaints, bronchitis, troublesome cough, etc. It is usually administered in the form of a syrup, which may be prepared by mixing any given quantity of the compound fluid white pine with three times its volume of simple syrup. For price of fluid extract see page 48; for price of syrup see page 176.

105. Yerba Santa, Aromatic. Aromatic Syrup of Yerba Santa is well known as a palatable vehicle for the administration of quinine and other bitter substances not acid in their nature. Each fluidounce will successfully mask the taste of 30 grains of quinine, held in suspension. The mixture should be dispensed with a shake label. We offer the syrup (see page 176) and the fluid extract (see page 50) for preparing the syrup extemporaneously. Two ounces of the fluid extract, when diluted, are sufficient to make one pint of the syrup.

106. Zymole. No liquid antiseptic on the market is superior to this in effectiveness, and but few approach it in elegance. It is adapted to internal and external use, and possesses a wide range of application, both as a medicinal and sanitary agent and a delightful toilet accessory. For price see page 205.

107. Solid Extracts, Assayed. Our solid extracts which contain alkaloids, or are otherwise of determinable strength, are carefully assayed and made to conform to specified standards. Each lot of extract is therefore of uniform medicinal strength with every other lot of the same kind. Owing to the very considerable variation in the yield of extractive from different samples of the same kind of drug, this is an item of great importance and should not be overlooked. See page 51.

TABLES OF INFORMATION.

A PRACTICAL SCHEME FOR URINE ANALYSIS.

The specimen of urine for examination should be taken from the whole quantity (well mixed) passed in 24 hours—that passed, say, from 7 o'clock one morning to the same hour the following morning, and should be tested before decomposition sets in. The principal points of note are enumerated in the following :

QUANTITY.—The amount voided in 24 hours varies, normally, within wide limits. The normal amount is usually stated as 1500 Cc. (50 fl. ozs.).

Increased—By excessive ingestion of liquids, as water, beer, milk, etc.; by cold and damp weather and other conditions interrupting cutaneous transpiration ; in diabetes, hysteria, in contracted kidney and waxy disease of the kidney.

Decreased—By ingestion of small quantity of liquids; in hot, dry weather (excessive perspiration); in diarrhea; in febrile conditions; in the earlier stages of dropsies; in certain forms of Bright's disease.

COLOR.—The color of normal urine is that of amber or slightly reddish-yellow. The color of urine is much affected by food and medicine, as well as by many morbid conditions. Santonin colors it a bright yellow; pyoctanin, a blue; madder, logwood, rhubarb, a red or brownish-red; strong coffee, turpentine, creosote, carbolic acid, etc., render it dark. A red or smoky color may indicate blood; greenish-yellow or brown, bile; whitish or turbid, earthy phosphates (excess), pus, or mucus.

Decreased quantity of urine is usually accompanied by increase in color, as in febrile conditions, etc., and, *vice versa*, as in diabetes, hysteria, etc.

REACTION.—Determined by the use of litmus-paper—acid turns the blue, red; and alkaline, the red, blue. The normal reaction of the urine is *slightly acid*, though at times it may be neutral. Nitrogenous food increases acidity; vegetable food decreases it. Some drugs influence the reaction of the urine, e. g., organic acids (citric, tartaric, etc.), and their salts with the alkali bases render the urine less acid. Mineral acids render the acidity greater. Alkalinity is generally due to decomposition of urea into ammonium carbonate, seen in retention, cystitis, etc. In this case, gentle heat will restore the red color of the litmus paper used.

ODOR.—Normally, characteristic, urinous. Concentration increases odor. Many foods, as asparagus, and drugs, as cubeb, turpentine, etc., greatly influence the odor. Turpentine gives the odor of violets. In diabetes mellitus, it is fragrant; in cystitis, retention, etc., with decomposition, it is fetid.

SPECIFIC GRAVITY.—Most conveniently ascertained by means of a urinometer. The urine should be at or very near the temperature at which the urinometer was graduated (generally 60° F.), and in a vessel sufficiently large to permit the instrument floating free of the sides. Read the sp. gr. from the top of the meniscus, on a level with the eye. The sp. gr. of normal urine is 1015 to 1025. In infants it is low—1007 to 1012. By multiplying the last two figures of the sp. gr. by 2.33 (Haser's coefficient) a close approximation to the number of parts, per 1,000, of solids contained may be obtained, and from this the total amount of solids passed in 24 hours may be easily calculated, which is normally 53 to 67 grammes (800 to 1,025 grains).

Increased—In diabetes mellitus; in cyanotic induration of the kidney; in acute diffuse nephritis; in concentrated urine, etc.

Decreased—In diabetes insipidus; in Bright's disease; by fasting; in copious quantity of urine, etc.

AVERAGE COMPOSITION OF NORMAL ADULT URINE (LEHMAN).

Water 932.019
Solid matter 67.981
　　　　Urea 32.909
　　　　Uric acid 1.098
　　　　Lactic acid 1.513
　　　　Lactates 1.732
　　　　Water extract 632
　　　　Spirit and Alcohol extract . . . 10.872
　　　　Sodium chloride, ⎫
　　　　Ammonium chloride, ⎬ . . . 3.712
　　　　Alkaline sulphates 7.321
　　　　Sodium phosphate 3.989
　　　　Calcium and Magnesium phos-
　　　　　　phates 1.108
　　　　Mucus 0.110

VARIATION IN QUANTITY OF NORMAL CONSTITUENTS.

URIC ACID—Tests.—A roughly approximate test of the diminution or increase of uric acid in urine may be applied as follows: Place in a test-tube f ℥ ij of the urine to be examined and the same quantity of normal urine in another test-tube. Now acidulate each with hydrochloric acid and set aside for 24 hours. A comparison of the two sediments (uric acid) will show the relative amounts.

Increased—Excess nitrogenous food; excessive tissue waste; diminished oxidation; gout; rheumatism; general malassimilation; diseases of the liver, etc.

Decreased—By vegetable diet; exercise; in chronic Bright's disease, etc.

UREA—Tests.—Place f ℥ ij of urine in a test-tube, add f ℥ ss of colorless nitric acid and set the tube in ice water. If urea is in excess, the characteristic crystals of urea nitrate will be precipitated. Increase of urea raises the sp. gr. of the urine. If the sp. gr. of the urine is lower than normal and no ppt. is obtained by the above test, evaporate f ℥ ij of the urine

to one-half its bulk, then apply the test as before. If
no ppt. (precipitate), the urea is below normal.

If there be no abnormal ingredients present, e, g.,
glucose, albumin, etc., the urea present may be
approximately estimated by calculating the amount of
solids present, as given under specific gravity (*vide*)
and dividing the result by 2.

Increased—In febrile conditions; by excess of
nitrogenous foods; in diabetes; epileptic attacks;
during administration of phosphorus, arsenic and
alcohol.

Decreased—In diseases of the liver; acute yellow
atrophy, carcinoma, etc.; in faulty excretion due to
renal disorders, biliary colic, etc.

CHLORIDES.—Place in a test-tube f3ij of the
urine to be examined, and in a companion tube the
same quantity of normal urine. Acidulate each with
nitric acid, add a solution of silver nitrate (1-50),
enough to ppt. the chlorides, set aside and let settle.
A comparison of the two ppts. will show variation
from normal.

Increased—By abundant drinking of water; increased ingestion of common salt; immediately following the crisis of acute febrile diseases, pneumonia, etc., if favorable termination is indicated.

Decreased—In diarrhea; in rapid formation of
large transudations; acute febrile conditions, especially
just prior to the crisis; acute and chronic diseases of
the kidneys with albuminuria; chronic diseases.

PHOSPHATES.—Put into a test-tube f3ij of the
urine; make decidedly alkaline with a few drops of
solution of caustic potash, boil and set aside for ppt.
to settle. At the end of 20 to 25 minutes the ppt.
should be about one-eighth the bulk of the urine, if
the quantity is normal.

Increased—By excess of nitrogenous food; in in-
flammatory diseases; mental strain; traumatic menin-

gitis; acute rheumatism; rickets; extensive bone disease. Decreased—In epilepsy; maniacal paroxysm; melancholia; general or sexual exhaustion.

NOTE.—The above simple tests are not given as yielding scientifically accurate results, but are only intended for rough approximation, which frequently will be of service to the physician in his clinical work. More elaborate and accurate tests will be found in any of the reliable works on urine analysis.

TESTS FOR ABNORMAL CONSTITUENTS.

ALBUMIN.—Place in a test-tube f ℥ ij of urine and heat to boiling. Add 10 or 12 drops of nitric acid. A small amount of albumin is shown by a slight diffuse cloudiness, a larger amount by a more or less flaky deposit, if considerable quantity, a firm mass will be formed.

Heat Test.—Filter the urine if necessary. The urine must be slightly acid; if not already so, add a drop or two of acetic acid. Now boil some in a test-tube. The presence of albumin will be indicated by an opalescence, a cloudiness, or a ppt

NOTE.—Urine containing pus or blood always contains albumin.

Picric Acid Test.—Picric acid has the advantage of being a test both for albumin and glucose (sugar), and its application in both is here given.

In a test-tube add equal volumes of the urine to be tested and a saturated aqueous solution of picric acid. The albumin, if present, is coagulated and is shown as a turbidity or a flaky precipitate, or a heavy mass owing to its quantity.

If there is no albumin, add sufficient solution of caustic potash to make the mixture alkaline and boil. If sugar is present, the mixture will be colored a dark red or brown, or black, the color dependent on the amount of sugar present.

GLUCOSE (SUGAR).—Picric Acid Test.—See under albumin.

Trommer's Test.—To a small quantity of urine in a test-tube, add a small amount of solution of copper sulphate, being careful not to get an excess of the latter; make strongly alkaline with solution of caustic potash and boil. A yellow or red ppt. indicates sugar.

Moore's Test.—Add to the urine about one-fourth its volume of caustic potash solution and apply heat. If glucose be present, the color of the mixture will become dark yellow or brown, and an odor of molasses will be evolved. Nitric acid added will more or less completely destroy the color.

BILIARY MATTERS.—Biliary coloring matters occur in the urine in different forms of icterus, and color the urine yellowish-brown, deep-brown, greenish-yellow, or nearly pure green. The foam produced by shaking the urine possesses a yellow or greenish tint.

Gmelin's Test.—Rosenbach's Modification.—Filter the urine through a very small filter. Apply to the filter, after the urine has all passed through, a drop of yellow nitric acid. A pale yellow spot will be formed, surrounded by a play of colors—red, violet, blue and green.

PUS.—The best means of detecting pus is the microscope, but Donne's pus-test may be applied. Let the urine stand in a test-tube, or, better, a conical glass, until the sediment is well settled; then carefully pour off the supernatant liquid; now add to the sediment a few drops of a strong solution of caustic potash and stir. A thick, slimy, tough mass will be formed.

BLOOD.—Guaiacum Test.—Place in a test-tube equal volumes of tinct. guaiacum and old turpentine which has been exposed to the action of the air under the influence of light for some time, and hence has absorbed oxygen. This mixture should not have the slightest blue color. Now cautiously add the urine to be tested. If blood be present, a ring will be formed at the union of the two liquids—changing from a bluish green to a blue color. Pus in the urine gives the same reaction, but can be differentiated from blood in that the former does not require the ozonized turpentine to obtain the reaction. The blue produced by pus is dissipated by heat, that by blood, not.

RULE FOR PROPORTIONING DOSES.
(DR. YOUNG'S).

To arrive at the proportionate dose for a person under adult age, add 12 to the age and divide the age by the result. Thus, for a child 2 years old the dose is 1-7 that of the adult dose:

$$\frac{2}{2+12} = \frac{1}{7}$$

Children do not tolerate narcotics well. Therefore, when administering them, only a little more than half the proportion indicated by the rule should be given. On the other hand, mild cathartics may be given in doses two or even three times the proportion. For hypodermic injections the dose should be only two-thirds to three-fourths of that given by mouth. When administered by rectum, about five-fourths of the amount can be administered.

RULES FOR CALCULATING THERMOMETRIC EQUIVALENTS.

ABOVE FREEZING POINT OF WATER, 32° F. (0°C.)

*To convert C. to F.—Multiply by 9, divide by 5, and add 32.

*To convert F. to C.—Subtract 32, multiply by 5, and divide by 9.

BELOW FREEZING POINT OF WATER AND ABOVE ZERO, F. (—17.77° C.)

To convert C. to F.—Multiply by 9, divide by 5, and subtract from 32.

To convert F. to C.—Subtract from 32, multiply by 5, and divide by 9.—degrees below zero.

BELOW ZERO, F. (—17.77° C.)

To convert C. to F.—Multiply by 9, divide by 5, and substract 32.— degrees below zero.

To convert F. to C.—Add 32, multiply by 5, and divide by 9.— degrees below zero.

*If the algebraic signs are observed, this rule applies equally well in all cases.

DROPS IN A FLUIDRACHM.

(FROM THE TABLE COMPILED BY S. L. TALBOT.)

A table showing the number of drops in a fluidrachm, also the weight of one fluidrachm in grains and grammes for each of the preparations named :

NAME.	Drops in fluidrachm, 60 min.	Weight of fluidrachm.	
		In grains.	In grammes.
Acidum Aceticum.....	108	58	3.75
Aceticum Dilutum	68	55	3.56
Carbolicum	111	59	3.82
Hydrochloricum	70	65	4.21
Dilutum	60	56	3.62
Hydrocyanicum Dilutum	60	54	3.49
Nitricum.	102	77	4.98
Dilutum	60	58	3.62
Nitrohydrochloricum	76	66	4.27
Phosphoricum Dilutum	59	57	3.69
Sulphuricum	128	101	6.54
Aromaticum	146	53	3.43
Dilutum	60	58.5	3.79
Alcohol	146	44	2.85
Dilutum	137	49	3.17
Aqua	60	55	3.56
Ammoniae Fortior	66	50	3.24
Destillata	60	53.5	3.46
Balsamum Peruvianum.	101	60	3.88
Chloroformum	260	80	5.18
Copaiba	410	51	3.30
Creosotum	122	56.5	3.66
Extractum Belladonnae Radicis Fluidum	156	57	3.69
Cinchonae Fluidum	138	59	3.75
Colchici Radicis Fluidum	160	57	3.69
Seminis Fluidum	158	55	3.56
Digitalis Fluidum	149	49	3.14
Gelsemii Fluidum	149	49	3.14
Hyoscyami Fluidum	160	59	3.82
Ipecacuanhae Fluidum	120	60	3.88
Rhei Fluidum	158	61	3.95
Valerianae Fluidum	150	49	3.17
Veratri Viridis Fluidum	150	50	3.24
Zingiberis Fluidum	142	48	3.11

NAME.	Drops in fluidrachm, 60 min.	Weight of fluidrachm.	
		In grains.	In grammes.
Glycerinum	67	68	4.40
Liquor Acidi Arsenosi	57	55	3.56
Arseni et Hydrargyri Iodidi	58	55	3.56
Ferri Citratis	71	72	4.66
Iodi Compositus	63	59	3.82
Potassae	62	58	3.75
Potassii Arsenitis	57	55	3.56
Oleoresina Aspidii	130	52	3.36
Oleum Amygdalae Amarae	115	55	3.56
Expressum	108	48.5	3.14
Anisi	119	54	3.49
Caryophylli	130	57	3.69
Cinnamomi	126	53.5	3.46
Cubebae	125	51	3.30
Gaultheriae	125	62	4.01
Limonis	129	47	3.04
Menthae Piperitae	129	50	3.24
Ricini	77	51.5	3.33
Sassafras	133	58	3.75
Terebinthinae	136	45.5	2.94
Tiglii	104	50	3.24
Spiritus Aetheris Compositus	148	45	2.91
Nitrosi	146	47	3.04
Ammoniae Aromaticus	142	48	3.11
Menthae Piperitae	142	47	3.04
Syrupus	65	72	4.66
Ferri Iodidi	65	77	4.98
Tinctura Aconiti	146	46	2.98
Belladonnae Foliorum	137	53	3.43
Cantharidis	131	51	3.33
Cinchonae Composita	140	49	3.17
Digitalis	128	53	3.43
Ferri Chloridi	150	53	3.43
Iodi	148	47	3.04
Nucis Vomicae	140	44	2.85
Opii	130	53	3.43
Camphorata	130	52	3.36
Deodorati	110	54	3.49
Valerianae	130	52	3.36
Veratri Viridis	145	46	2.98
Zingiberis	144	46	2.98
Vinum Colchici Radicis	107	55	3.56
Seminis	111	54	3.49
Opii	100	55	3.56

OFFICIAL RULES FOR DILUTING ALCOHOL.
(U. S. PHARMACOPŒIA).

I. *By Volume.*—Designate the volume-percentage of the stronger alcohol by V, and that of the weaker alcohol by v. *Rule:* Mix v volumes of the stronger alcohol with pure water to make V volumes of product. Allow the mixture to stand until full contraction has taken place and until it has cooled; then make up any deficiency in the V volumes by adding more water. *Example:* An alcohol of 30 per cent. by volume is to be made from an alcohol of 94 per cent. by volume. Take 30 volumes of the 94 per cent. alcohol and add enough water to produce 94 volumes.

II. *By Weight.*—Designate the weight-percentage of the stronger alcohol by W, and that of the weaker by w. *Rule:* Mix w parts by weight of the strong alcohol with pure water to make W parts by weight of product. *Example:* An alcohol of 50 per cent. by weight is to be made from an alcohol of 91 per cent. by weight. Take 50 parts by weight of the 91 per cent. alcohol and add enough pure water to produce 91 parts by weight.

ALCOHOL DILUTION TABLE.
(UNOFFICIAL).

To make alcohol of	Take of 95 per cent. alcohol	And mix with pure water
90 per cent.	18 volumes	1 volume.
85 "	17 "	2 volumes.
80 "	16 "	3 "
75 "	15 "	4 "
70 "	14 "	5 "
65 "	13 "	6 "
60 "	12 "	7 "
55 "	11 "	8 "
50 "	10 "	9 "
45 "	9 "	10 "
40 "	8 "	11 "
35 "	7 "	12 "
30 "	6 "	13 "
25 "	5 "	14 "
20 "	4 "	15 "
15 "	3 "	16 "
10 "	2 "	17 "
5 "	1 volume	18 "

EQUIVALENT WEIGHTS· AND. MEASURES.

MEASURES.

	About	to	Exactly
1 minim	.06 Cc.		.06161 Cc.
1 fl. dr. ('60 minims.)	4. "		3.6967 "
1 fl. oz.	30. "		29.574 "
1 pint	.473. "		473.179 "
1 quart	1. liter		.946 liter
1 gallon	4. liters.		3.785 liters
1 Cc.	16 minims		16.23 m.
4 "	1 fl. dr.		64.92 m.
15 "	½ fl. oz.		243.46 m.
30 '	1 fl. oz.		1 fl. oz. 6.9 m.
250 "	8⅓ fl. ozs.		8 fl. ozs. 217.7 m.
500 "	16⅔ fl. ozs.		16 fl. ozs. 435.3 m.
1 liter	33 fl. ozs.		33 fl. ozs. 390.6 m.
4 liters	1 gallon		1 gal. 7 ozs. 122.4 m.

WEIGHTS.

	About	Exactly
1-100 grain	.00065 Gm.	.000648 Gm.
1-64 "	.001 Gm. (1 mgr.)	.001013 "
1-32 "	.002 Gm.	.002025 "
1-10 "	.0065 "	.006489 "
1-2 "	.032 "	.0324 "
1 "	.065 "	.06479 "
5 grains	.300 "	.324 "
15 "	1. "	.972 "
60 "	4. "	3.888 "
1 av. oz.	28. "	28.35 "
4 "	113. "	113.398 "
1 pound.	450. "	453.592 "
1 milligramme (.001 Gm.)	1-65 grain	0.015 grain.
10 mgrs. (.01 Gm.)	1-6 "	0.154 "
65 " (.065 Gm.)	1 "	1.0031 "
100 " (.100 Gm.)	1 1-2 grains	1.543 "
1 gramme	15 "	15.4324 "
4 grammes	60 "	61.729 "
10 "	1-3 oz.	154.324 "
28 "	1 oz.	432.107 "
100	3 1-2 ozs.	3 ozs. 230.7 grs.
500 "	1 1-10 lbs.	1 lb. 1 oz. 278.7 grs.
1000 " (1 kilo)	2 1-5 lbs.	2 lbs. 3 ozs. 119.9 grs.

PERCENTAGE SOLUTIONS.

The first table being for the weaker solutions, states the quantities of the ingredients to make up one pint of solution; the second table being for the stronger solutions, usually made up only in small amounts, is calculated for (approximately) one ounce of the finished solution. In most cases fractions of grains are rounded off, as by this means results fully accurate enough for all practical purposes can be had:

I.

Percentage strength.		For each 16 fl. ozs. of water take of the salt
0.020 per cent.	1 in 5000	1.4 grains.
0.100 "	1 " 1000	7.3 "
0.111 "	1 " 900	8. "
0.125 "	1 " 800	9. "
0.143 "	1 " 700	11. "
0.167 "	1 " 600	12. "
0.20 "	1 " 500	15. "
0.25 "	1 " 400	18. "
0.33 "	1 " 300	24. "
0.50 "	1 " 200	37. "

II.

Percentage strength.		For each fl. oz. of water take of the salt
1. per cent.	1 in 100	4.6 grains.
2. "	1 " 50	9. "
3. "	1 " 33	14. "
4. "	1 " 25	19. "
5. "	1 " 20	24. "
6. "	1 " 17	29. "
7. "	1 " 14	34. "
8. "	1 " 12	40. "
9. "	1 " 11	45. "
10. "	1 " 10	51. "
15. "	1 " 7	81. "
20. "	1 " 5	114. "
25. "	1 " 4	152. "

RELATIVE WEIGHTS AND VOLUMES OF A FEW IMPORTANT OFFICIAL LIQUIDS.

16 fluidounces weigh.			1 pound av. measures	
13 2-3 av. ozs.	Alcohol		18 3-4 fluid ounces	
16	" "	Ammonia water	16	" "
15	" "	Ammonia water, stronger	17	" "
16	" "	Castor oil	16	" "
24 3-4	" "	Chloroform	10 1-3	" "
12	" "	Ether	21 1-5	" "
20 3-4	" "	Glycerin	12 1-4	" "
19 3-4	" "	Goulard's extract	12 7-8	" "
19	" "	Hydrochloric acid	13 1-4	" "
25 1-2	" "	Monsel's solution	10	" "
23 3-4	" "	Nitric acid	10 3-4	" "
14	" "	Spirit of Nitrous Ether	18 1-4	" "
30 1-2	" "	Sulphuric acid	8 3-8	" "
21 4-5	" "	Syrup	11 2-3	" "
16 2-3	" "	Water	15 1-3	" "

POISONS AND THEIR ANTIDOTES.

(Era Dose Book).

GROUP I.

Acids.—Acetic, Hydrochloric, Nitric, Nitro-muriatic, Sulphuric.

TREATMENT.—Give no emetic. Give at once large draughts of water, (or milk) with chalk, whiting, magnesia, or baking soda, or give strong soapsuds to neutralize acid; olive oil, white of egg beaten up with water, and later, mucilaginous drinks of flaxseed or slippery elm, are useful. Give laudanum (20 drops) if much pain.

GROUP II.

Carbolic acid, Creosote, Resorcin.

TREATMENT.—Promote vomiting with warm water containing baking soda, or cause it with mustard (a tablespoonful stirred to a cream with water). Give white of egg beaten up

with water, or olive oil, (a cupful); stimulants, (whiskey, etc.)
freely; warmth and friction to the extremities.

GROUP III.

Antimony, Chromium, Copper, Iodine, Mercury, Zinc, their
compounds and preparations; Cantharides, Colchicum,
Elaterium and Croton, Savin and Tansy oils.

TREATMENT.—Give white of eggs (half dozen or more, raw)
or flour mixed with water. Promote vomiting with warm
water containing baking soda, or cause it with mustard (a
tablespoonful stirred to a cream with water). Give strong tea
or coffee, stimulants if needed, laudanum (20 drops) if much
pain; demulcent drinks of flaxseed or slippery elm.

GROUP IV.

Caustic Alkalies (Potash, Soda, Ammonia, etc.)

TREATMENT.—Promote vomiting by large draughts of warm
water. Give vinegar or diluted lemon juice; olive oil, the whites
of eggs, beaten up with water, gruel or demulcent drinks ot
flaxseed or slippery elm; laudanum (20 drops) if much pain.

GROUP V.

Alcohol, Benzin, Benzol, Camphor, Carbon bisulphide, Chloral,
Chloroform, Ether, Hydrocyanic acid, its compounds and
preparations.

TREATMENT.—If necessary, give emetic of mustard (a tablespoonful stirred to a cream with water). Let patient have
plenty of fresh air; maintain a horizontal position. Keep the
body warm but try to rouse the patient by ammonia to nostrils,
cold douche to the head, friction and mustard plasters to limbs,
etc. Use artificial respiration.

GROUP VI.

Cannabis Indica, Opium, Coca—their alkaloids, salts and preparations.

TREATMENT.—Give emetic (if necessary) of mustard (a
tablespoonful stirred to a cream with water); followed by large
draughts of warm water. Then strong tea or coffee. Rectal
injections of tincture of capsicum. Arouse the patient, and

keep him awake and in motion.' Keep up artificial respiration, even after life seems to be extinct.

GROUP VII.

Aconite, Digitalis, Lobelia, Tobacco, Veratrum album and Veratrum viride—their alkaloids, salts and preparations.

TREATMENT.—Give emetic of mustard (a tablespoonful stirred to a cream with water), followed by large draughts or warm water. Give strong tea or coffee, with powdered charcoal; stimulants (whiskey, etc.,) freely; warmth to the extremities; keep the patient in a horizontal position; use artificial respiration persistently.

GROUP VIII.

Belladonna, Calabar Bean, Conium, Gelsemium, Hyoscyamus, Santonin, Stramonium—their alkaloids, salts and preparations.

TREATMENT.—Give emetic of mustard (a tablespoonful stirred to a cream with water), followed by large draughts of warm water; give strong tea or coffee, with powdered charcoal; stimulants, (whiskey, etc.,) if necessary; rouse the patient if drowsy; heat and friction to extremities; artificial respiration.

GROUP IX.

Cocculus Indicus, Nux Vomica—their alkaloids, salts and preparations.

TREATMENT.—Give emetic of mustard (a tablespoonful stirred to a cream with water), followed by large draughts of warm water. Give powdered charcoal, iodide of starch, or tannin. To relieve spasms, let patient inhale pure chloroform, or give chloral hydrate (25 grains) or potassium bromide (½ oz).

GROUP X.

Arsenic and its compounds, (Cobalt), Paris Green, "Rough on Rats," etc.

TREATMENT.—Promote vomiting with warm water, or cause it with mustard (a tablespoonful stirred to a cream with water). Procure at once from the drug store, hydrated oxide of iron, and give a cupful of it (or mix a teaspoonful of calcined magnesia with a cup of water, add three tablespoonfuls of tincture of iron, mix well and give the whole of it). Follow with olive

oil, or white of eggs (raw) and mucilaginous drinks. Laudanum (20 drops) if much pain.

GROUP XI.

Oxalic acid and its soluble salts.

TREATMENT.—Give chalk or whiting (tablespoonful), or
even air-slacked lime (a teaspoonful in fine powder) mixed
with two tablespoonfuls of vinegar. (Do *not* give soda or
potash to neutralize the acid). Promote vomiting by large
draughts of water, or cause it with mustard (a tablespoonful
stirred to a cream with water). Give olive oil and mucilaginous drinks; stimulants, (whiskey, etc.,) and warmth to extremities.

GROUP XII.

Barium and its salts, Lead and its salts.

TREATMENT.—Give Epsom salt (½ oz.) or Glauber's salt
(one oz.) dissolved in a tumbler of water. Promote vomiting
by warm water, or cause it with mustard (a tablespoonful
stirred to a cream with water). Give milk, demulcent drinks
of flaxseed or slippery elm, and laudanum (20 drops) if much
pain.

GROUP XIII.

Silver nitrate. (Lunar Caustic).

TREATMENT.—Give common salt (a tablespoonful dissolved
in a tumblerful of warm water); then an emetic of mustard (a
tablespoonful stirred to a cream with water) followed by large
draughts of warm water. Later, give gruel, arrowroot, or
demulcent drinks of flaxseed or slippery elm.

GROUP XIV.

Phosphorus Compounds (Rat Paste, etc.)

TREATMENT.—Give an emetic of mustard (a tablespoonful
stirred to a cream with water), or better, of blue vitriol, 3
grains every five minutes, until vomiting occurs. Give a teaspoonful of old thick oil of turpentine; also Epsom salt (½ oz.
in a tumblerful of water). Do *not* give oil, except the turpentine.

SYNONYMS.

This list is intended to be used in connection with the fluid extract list, and names occurring in that list are not repeated in the alphabetical arrangement of this. In general, consult the fluid extract list for the more common names, and this list for the less common and the botanical titles. Official botanical names occur in italics. The names in the right hand column refer to those of the fluid extract list, under which are given the botanical name, properties, dose and price of each. Botanical synonyms are marked with an asterisk (*).

Abies balsamea	Balsam Fir.
Abies Canadensis*	Hemlock Spruce.
Acacia Catechu.	Catechu.
Achillea Millefolium	Yarrow.
Aconitum Napellus	Aconite.
Aconitum vulgare*	Aconite.
Acorus Calamus.	Calamus.
Actaea racemosa*	Black Cohosh.
Aegle Marmelos	Bael fruit.
Aesculus glabra	Buckeye.
Aesculus Hippocastanum	Horse chestnut.
Agave planifolia	Guaco.
Agrimonia Eupatoria	Agrimony.
Agrimonia striata*	Agrimony.
Agropyrum repens	Couch Grass.
Alkanna tinctoria	Alkanet.
Allium sativum	Garlic.
Alnus rugosa*	Tag Alder.
Alnus serrulata	Tag Alder.
Aloe Perryi	Aloes, Socotrine.
Alpina Galanga*	Galanga.
Alpinia officinarum	Galanga.
Althaea officinalis	Marshmallow.
American Gentian	Five-flowered Gentian.
American valerian	Ladies' Slipper.
Amomum repens*	Cardamom.
Ampelopsis quinquefolia	American Ivy.
Amygdalus Persica*	Peach tree.
Anacyclus Pyrethrum	Pellitory.
Anagallis arvensis	Scarlet Pimpernel.
Anamirta Cocculus*	Cocculus Indicus.
Anamirta paniculata	Cocculus Indicus.
Anchusa tinctoria*	Alkanet.
Andira inermis	Cabbage Tree.

Andromeda arborea* . . . . . . Sourwood.
Andropogon saccharatus*. . . . Broom Corn.
Anemone Hepatica*. . . . . . . Liverwort.
Anemone pratensis. . . . . . . Pulsatilla.
Anemone Pulsatilla . . . . . . Pulsatilla.
Anethum graveolens. . . . . . . Dill.
Anthemis nobilis. . . . . Chamomile, Roman.
Apium graveolens. . . . . Celery-seed.
Apium Petroselinum* . . . . . . Parsley.
Aplopappus Baylahuen . . . . . Hysterionica.
Apocynum androsaemifolium . . Bitter root.
Apocynum cannabinum . . . . Canadian Hemp.
Apple Peru. . . . . Stramonium.
Aralia hispida . . . . Dwarf Elder.
Aralia nudicaulis . . . . American Sarsaparilla.
Aralia racemosa . . . . Spikenard.
Arbutus Uva Ursi*. . . . . . . Uva Ursi.
Archangelica atropurpurea* . . Angelica seed.
Archangelica officinalis* . . . . Angelica root.
Arctium Lappa . . . . Burdock.
Arctium majus* . . . . Burdock.
Arctostaphylos glauca . . . . . Manzanito.
Arctostaphylos Uva Ursi . . . Uva Ursi.
Arisaema triphyllum . . . . . . Indian Turnip.
Aristolochia reticulata . . . . Serpentaria.
Aristolochia Serpentaria . . . Serpentaria.
Artanthe elongata* . . . . . . Matico.
Artemisia Absinthium . . . . Wormwood.
Artemisia frigida . . . . Mountain Sage.
Artemisia maritima* . . . . . . Levant Wormseed.
Artemisia pauciflora . . . . . Levant Wormseed.
Artemisia vulgaris . . . . Mugwort.
Arum triphyllum* . . . . Indian Turnip.
Asagraea officinalis . . . . . Cevadilla seed.
Asarum Canadense. . . . . . . Canada Snake root.
Asclepias Cornuti* . . . . Silkweed.
Asclepias curassavica. . . . . . Blood-flower.
Asclepias incarnata . . . . . . Swamp Milkweed.
Asclepias Syriaca . . . . Silkweed.
Asclepias tuberosa . . . . . . Pleurisy-root.
Asimina triloba. . . . . Papaw seed.
Aspen . . . . White Poplar.
Aspidium Filix-mas* . . . . . . Male-fernp
Aspidium marginale* . . . . . . Male-fern.
Astringent root. . . . . Cranesbill.
Atropa Belladonna . . . . . . Belladonna.
Ava Kava . . . . Kava kava
Avens . . . . Water Avens.
Balsamodendron Myrrha* . . . Myrrh.
Balsam Spruce. . . . . Balsam Fir.
Balsam Tolu . . . . Tolu.
Banksia Abyssinica* . . . . . . Kousso.
Baptisia tinctoria. . . . . Wild Indigo.

Barosma betulina	Buchu.
Barosma crenulata	Buchu.
Bean of St. Ignatius	Ignatia.
Bearberry	Uva Ursi.
Bearsbed	Haircap Moss
Bedstraw	Cleavers.
Bengal Quince	Bael fruit.
Berberis nervosa*	Berberis Aquifolium.
Berberis repens*	Berberis Aquifolium.
Berberis vulgaris	Barberry.
Betonica officinalis*	Wood Betony.
Betula rugosa.	Tag Alder.
Bhang	Cannabis Indica.
Bicuculla Canadensis*	Corydalis.
Bidens bipinnata	Spanish needles.
Bignonia Caroba*	Caroba.
Bignonia sempervirens*	Gelsemium.
Birthroot	Beth-root.
Bitter apple	Colocynth.
Bitter cucumber	Colocynth.
Bitterstick	Chirata.
Bitter thistle	Blessed Thistle.
Black cherry	Belladonna.
Black Indian Hemp	Canadian Hemp.
Black larch.	Tamarac.
Black root	Culver's root.
Black Sampson	Echinacea.
Black Snakeroot	Black Cohosh.
Blooming spurge	Flowering Spurge.
Blue Bells	Abscess root.
Bogbean	Buckbean.
Boldoa fragrans*	Boldo.
Boletus laricis*	Agaric, White.
Bombay root	Galanga.
Bouncing Bet	Soapwort.
Boxwood	Dogwood.
Brayera anthelmintica* . . .	Kousso.
Brittlestem	Dwarf Elder.
Broad-leaved Laurel	Mountain Laurel
Brookbean.	Buckbean.
Broom flowers	Broom.
Brunfelsia Hopeana	Manacá.
Bullsfoot	Coltsfoot.
Burning Bush	Wahoo.
Bursa Bursa-Pastoris* . . .	Shepherd's-purse.
Butterfly-weed	Pleurisy-root.
Calico Bush	Mountain Laurel
Callicocca Ipecacuanha* . .	Ipecac.
California Poppy	Eschscholtzia Californica.
Camellia Thea	Tea.
Canada Fleabane	Fleabane.
Cancer-root	Beech-drop.
Candleberry	Bayberry.

Cannabis sativa . . . . Cannabis Indica.
Capsella Bursa-pastoris . . . . . Shepherd's-purse.
Carbenia benedicta* . . . . Blessed Thistle.
Carthamus tinctorius . . . . . . American Saffron.
Carum Carvi . . . . Caraway.
Carum Petroselinum* . . . . . . Parsley.
Caryophyllus aromaticus . . . . Cloves.
Cassia acutifolia . . . . Senna.
Cassia angustifolia . . . . Senna.
Cassia elongata* . . . . Senna.
Cassia lanceolata* . . . . Senna.
Castalia odorata* . . . . White Pond Lily.
Castanea dentata ! . . . . Chestnut.
Castanea vesca* . . . . Chestnut.
Castella Nicholsoni . . . . Chapparro Amargoso.
Catarrh root . . . . Galanga.
Catchfly . . . . Bitter-root.
Catchweed . . . . Cleavers.
Catsmint . . . . Catnep.
Catswort . . . . Catnep.
Cayenne pepper . . . . Capsicum.
Ceanothus Americanus . . . . . Jersey Tea.
Celastrus scandens . . . . False Bittersweet.
Centaurea benedicta* . . . . . . Blessed Thistle.
Cephaëlis Ipecacuanha . . . . . Ipecac.
Cephalanthus occidentalis . . . Button Bush.
Cerasus serotina* . . . . Wild Cherry.
Cercis Canadensis . . . . Judas Tree.
Cervispina cathartica* . . . . . Buckthorn berries.
Chamaelirium Carolinianum* . . False Unicorn.
Chamaelirium luteum . . . . . . False Unicorn.
Chamomilla officinalis* . . . . . Chamomile, German.
Checkerberry . . . . Squaw Vine.
Checkerberry . . . . Wintergreen.
Chelone alba* . . . . Balmony.
Chelone glabra . . . . Balmony.
Chequen . . . . Chekan.
Chickentoe . . . . Crawley root.
Chinchirocoma . . . . Mutisia viciaefolia.
Chinese sumach . . . . Ailanthus glandulosa.
Chionanthus Virginica . . . . . Fringetree.
Chironia angularis* . . . . . . American Centaury.
Chittem bark . . . . Cascara Sagrada.
Chocolate root . . . . Water Avens.
Chondodendron tomentosum . . Pareira Brava.
Christmas Rose . . . . Black Hellebore.
Chrysanthemum Chamomilla* . . Chamomile, German.
Chrysanthemum Leucanthemum.Ox-Eye Daisy.
Chrysanthemum Parthenium* . . Feverfew.
Churrus . . . . Cannabis Indica.
Cichorium Intybus . . . . Chicory.
Cicuta maculata . . . . Water Hemlock.
Cinnamomum Cassia . . . . . . Cassia Cinnamon.

Cinnamomum Zeylanicum	Ceylon Cinnamon.
Citrullus Colocynthis	Colocynth.
Citrullus vulgaris	Watermelon.
Citrus Aurantium	Orange peel, sweet.
Citrus Limonum	Lemon peel.
Citrus vulgaris	Orange peel, bitter.
Claviceps purpurea	Ergot.
Climbing Bittersweet	False Bittersweet.
Clove Garlic	Garlic.
Cnicus benedictus	Blessed Thistle.
Coakum	Poke.
Cocculus palmatus*	Columbo.
Coccus cacti	Cochineal.
Cochlearia Armoracia.	Horse-radish.
Cola acuminata	Kola.
Colic root	Wild Yam.
Collinsonia Canadensis	Stone root.
Commiphora Myrrha	Myrrh.
Comptonia asplenifolia	Sweet Fern.
Consumptive's weed	Yerba Santa.
Convallaria biflora*	Solomon's Seal.
Convallaria majalis	Lily of the Valley.
Convolvulus Purga*	Jalap.
Coptis trifolia	Gold Thread.
Corallorhiza odontorhiza	Crawley root.
Coral root	Crawley root.
Cordiceps purpurea*	Ergot.
Corn Smut	Corn Ergot.
Cornus circinata	Green Osier.
Cornus Florida	Dogwood.
Cornus rugosa*	Green Osier.
Cornus sericea	Red Osier.
Coumarouma odorata*	Tonka Bean.
Cow Parsnip	Masterwort.
Crataeva Marmelos*	Bael fruit.
Crataeva religiosa*	Bael fruit.
Croton Eluteria	Cascarilla.
Croton Philippinensis*	Kamala.
Cubeba officinalis*	Cubeb.
Cucumis Citrullus	Watermelon.
Cucumis Colocynthis*	Colocynth.
Cucurbita Citrullus*	Watermelon.
Cucurbita Pepo	Pumpkin.
Culver's Physic	Culver's root.
Cunila pulegioides*	Pennyroyal.
Curcuma longa	Turmeric.
Curcuma rotunda*	Turmeric.
Curcuma Zedoaria	Zedoary.
Curled Dock	Yellow Dock.
Cusparia febrifuga*	Angustura.
Cusparia trifoliata*	Angustura.
Cutch	Catechu.
Cymbidium Odontorhizon*	Crawley root.

Cyperus articulatus . . . . Adrue.
Cypripedium parviflorum . . . Ladies' Slipper.
Cypripedium pubescens . . . . Ladies' Slipper.
Cytisus Scoparius . . . . . Broom.
Daphne Mezereum . . . . . Mezereum.
Datura Stramonium . . . . . Stramonium.
Deadly Nightshade . . . . . . Belladonna.
Deadnettle . . . . Angelica.
Deerberry . . . . Wintergreen.
Delphinium consolida . . . . Larkspur.
Delphinium Staphisagria . . Stavesacre.
Dicentra Canadensis . . . . . Corydalis.
Diclytra Canadensis* . . . . Corydalis.
Dioscorea villosa . . . . . . Wild Yam.
Diospyros Virginiana . . . . Persimmon.
Dipteryx odorata . . . . . . Tonka Bean.
Dipteryx oppositifolia . . . Tonka Bean.
Ditch Stonecrop . . . . . . Virginia Stonecrop.
Dog-grass . . . . Couch-grass.
Dogsbane . . . . Bitter-root.
Dracontium foetidum* . . . . Skunk Cabbage.
Dragon root . . . . Indian Turnip.
Drosera rotundifolia . . . . Sundew.
Dryopteris Filix-mas . . . Male Fern.
Dryopteris marginalis . . . Male Fern.
Dwale . . . . Belladonna.
Dyer's Oak . . . . Galls.
Dyer's Saffron . . . . . . American Saffron
Elettaria repens . . . . . Cardamom.
Emetic herb . . . . Lobelia.
Emetic root . . . . Flowering Spurge.
Epigea repens . . . . Gravel plant.
Epilobium angustifolium . . . Willow herb.
Epilobium spicatum* . . . . Willow herb.
Epiphegus Virginiana . . . . Beech-drop.
Equisetum hyemale . . . . . Scouring Rush.
Erechthites hieracifolia . . . Fireweed.
Erigeron Canadense . . . . Fleabane.
Eryngium aquaticum . . . . . Water Eryngo.
Eryngium yuccaefolium* . . . Water Eryngo.
Erythroxylon Coca . . . . Coca.
Eucalyptus globulus . . . . Eucalyptus.
Eucalyptus rostrata . . . . . Red Gum.
Eugenia aromatica . . . . Cloves.
Eugenia caryophyllata* . . . Cloves.
Eugenia Chekan . . . . . . Chekan.
Eugenia Jambolana . . . . . Jambul seed.
Eugenia Pimenta* . . . . . Pimenta.
Eupatorium perfoliatum . . Boneset.
Eupatorium purpureum . . . . Queen of the Meadow
Euphorbia corollata . . . . Flowering Spurge.
Euphorbia pilulifera . . . . Euphorbia.
Euryangium Sumbul* . . . . Musk-root.

Exogonium Purga*	Jalap.
Fabiana imbricata	Pichi.
False valerian	Life-root.
False white cedar	Arbor vitæ.
Ferula fœtida	Asafetida.
Ferula Sumbul	Musk-root.
Fever-bush	Spice-bush.
Fieldbalm	Catnep.
Five-leaved Ivy	American Ivy.
Flag Lily	Blue Flag.
Flytrap	Bitter root.
Foeniculum capillaceum	Fennel.
Foeniculum vulgare*	Fennel.
Foreign Indian Hemp	Cannabis Indica.
Franciscea uniflora*	Manaca.
Frangula vulgaris*	Buckthorn bark.
Frankenia grandifolia	Yerba Reuma.
Frasera Carolinensis	American Columbo.
Frasera Walteri*	American Columbo.
Fraxinus Americana	White Ash, American.
Fraxinus sambucifolia*	Black Ash.
Fucus vesiculosus	Bladder wrack.
Galipea Cusparia	Angustura.
Galipea officinalis*	Angustura.
Galium Aparine	Cleavers.
Galium verum	Ladies' Bed-straw.
Garcinia Mangostana	Mangosteen.
Garget	Poke.
Garrya Fremontii	California Fever-bush.
Gaultheria procumbens	Wintergreen.
Gayfeather	Button Snakeroot.
Gelsemium nitidum*	Gelsemium.
Gentiana lutea	Gentian.
Gentiana quinqueflora	Five-flowered Gentian.
Gentiana quinquefolia*	Five-flowered Gentian.
Geum rivale	Water Avens.
Gillenia trifoliata	Indian Physic.
Gnaphalium obtusifolium*	Life Everlasting.
Gnaphalium polycephalum	Life Everlasting.
Golden senecio	Life-root.
Gonolobus Condurango	Condurango.
Gossypium herbaceum	Cotton-root bark.
Gouania Domingensis	Chewstick.
Granatum	Pomegranate root bark.
Gravel root	Queen of the Meadow.
Ground Laurel	Gravel plant.
Ground Lily	Beth root.
Guaiacum officinale	Guaiac wood and resin.
Guaiacum sanctum	Guaiac wood.
Guarea Rusbyi	Cocillana.
Gulfweed	Bladder wrack.
Gum Benjamin	Benzoin.
Gunjah	Cannabis Indica.

Hackmatac	Tamarac.
Hæmatoxylon Campechianum	Logwood.
Hagenia Abyssinica	Kousso.
Haplopappus Baylahuen	Hysterionica.
Hashish	Cannabis Indica.
Hedera quinquefolia*	American Ivy.
Helianthemum Canadense	Frostwort.
Helianthus annuus	Sunflower.
Hellebore, white	White Hellebore.
Helleborus niger	Black Hellebore.
Helleborus trifolia*	Gold Thread.
Helonias dioica*	False Unicorn.
Helonias viride*	Veratrum viride.
Hepatica triloba	Liverwort.
Heracleum lanatum	Masterwort.
High cranberry	Viburnum Opulus.
Holy thistle	Blessed thistle.
Hoodwort	Scullcap.
Hoptree	Wafer Ash.
Horsefly-weed	Wild Indigo.
Horseweed	Stone-root.
Huaco	Guaco.
Humulus Lupulus	Hops and Lupulin
Hypericum perforatum	Johnswort.
Ignatiana amara*	Ignatia.
Ignatiana Philippinica*	Ignatia.
Ilex Paraguayensis	Paraguay Tea.
Ilex verticillata	Black Alder.
Indian Senna	Senna.
Indian Arrow	Wahoo.
Indian Cannabis	Cannabis Indica.
Indian Ginger	Canada Snakeroot.
Indian Hemp, black	Canadian Hemp.
Indian Hemp, foreign	Cannabis Indica.
Indian Hemp, white	Swamp Milkweed.
Indian Lettuce	American Columbo.
Indian Paint	Blood-root.
Indian Sage	Boneset.
Indian Tobacco	Lobelia.
Inula Helenium	Elecampane.
Ipomoea Jalapa	Jalap.
Ipomoea Purga*	Jalap.
Iris Florentina	Orris root.
Iris versicolor	Blue Flag.
Itch-weed	Veratrum Viride.
Jacaranda procera	Caroba.
Jack in the Pulpit	Indian Turnip.
Jamestown-weed	Stramonium.
Jateorhiza Calumba*	Columbo.
Iateorhiza palmata	Columbo.
Jeffersonia diphylla	Twin leaf.
Jesuit's bark	Cinchona.
Jimsonweed	Stramonium.

Mallotus Philippinensis . . . Kamala. .
Mango. . . . . Mangosteen.
Marigold . . . . Calendula.
Marsh Clover . . . . Buckbean.
Matricaria Chamomilla . . . . Chamomile, German.
Matricaria Parthenium* . . . Feverfew.
May apple . . . . Mandrake.
Maypops . . . . Passiflora incarnata.
Meadow Anemone . . . . . . Pulsatilla. .
Meadowpride. . . . . American Columbo.
Meadow Saffron . . . . Colchicum. .
Melia Azedarach . . . . Pride of China. .
Menispermum Canadense . . Yellow Parilla.
Menispermum Cocculus* . . . Cocculus Indicus.
Menispermum Virginicum* . . Yellow Parilla.
Mentha piperita . . . . . . Peppermint.
Mentha spicata* . . . . Spearmint.
Mentha viridis . . . . . . Spearmint.
Menyanthes trifoliata . . . . . Buckbean.
Mezereum officinarum* . . . . Mezereum.
Micromeria Douglasii . . . . Yerba Buena.
Milfoil. . . . . Yarrow.
Mimosa Catechu* . . . . . . Catechu.
Monarda fistulosa . . . . . . Wild Bergamot.
Monarda mollis* . . . . Wild Bergamot.
Monkshood . . . . Aconite.
Moonseed . . . . Yellow Parilla.
Mountain balm . . . . Yerba Santa.
Mountain grape . . . . Berberis Aquifolium.
Mountain rush . . . . Ephedra.
Mountain tea . . . . Wintergreen.
Mountain tobabacco . . . . . Arnica.
Mouth-root. . . . . Gold Thread.
Murillo bark . . . . Soap-tree bark.
Myrcia acris . . . . Bay Laurel.
Myrica asplenifolia . . . . . Sweet Fern.
Myrica cerifera. . . . . Bayberry.
Myristica aromatica* . . . . . Mace and Nutmeg.
Myristica fragrans . . . . . Mace and Nutmeg.
Myristica moschata* . . . . . Mace and Nutmeg.
Myristica officinalis* . . . . . Mace and Nutmeg.
Myrospermum toluiferum* . . Tolu.
Myroxylon toluifera* . . . . Tolu.
Myrtleflag . . . . Calamus.
Myrtus acris* . . . . Bay Laurel.
Myrtus Chekan* . . . . . . Chekan.
Naked Ladies . . . . Colchicum.
Narrow Dock . . . . Yellow Dock.
Nasturtium Armoracia* . . . Horse Radish.
Nectandra . . . . Coto.
Nepeta Cataria . . . . Catnep, Catnip.
Nicotiana Tabacum . . . . Tobacco.
Nosebleed . . . . Yarrow.

Nuphar advena	Yellow Pond-lily.
Nymphæa odorata	White Pond-lily.
Oenanthe Phellandrium	Water Fennel.
Oenothera biennis	Evening Primrose.
Old Man's Beard	Fringetree bark.
Olive Spurge	Mezereum.
Onagra biennis*	Evening Primrose.
Ophelia Chirata*	Chirata.
Opium Poppy	Poppy.
Ordeal bean	Calabar Beau.
Oregon grape	Berberis Aquifolium.
Osmunda regalis	Buckhorn brake.
Ostrya Virginica	Iron-wood.
Oxydendron arboreum	Sourwood leaves.
Paeonia officinalis	Peony.
Panama bark	Soaptree bark.
Papaver somniferum	Poppy head, Opium.
Pappoose root	Blue Cohosh.
Parthenocissus quinquefolia* . .	American Ivy.
Partridgeberry	Squaw-vine.
Partridgeberry	Wintergreen.
Paullinia Cupana	Guarana.
Paullinia sorbilis*	Guarana.
Peachwood	Logwood.
Penthorum sedoides	Virginia Stonecrop.
Persica vulgaris*	Peach tree.
Peruvian bark	Cinchona.
Petroselinum sativum	Parsley.
Peumus Boldus	Boldo.
Peumus fragrans*	Boldo.
Phoradendron flavescens . . .	Mistletoe.
Pickpocket	Shepherd's Purse.
Picraena excelsa	Quassia.
Picramnia (sp. undetermined)	Cascara Amarga.
Pigeonberry	Poke.
Pilocarpus Jaborandi	Jaborandi.
Pilocarpus Sellodnus	Jaborandi.
Pimenta officinalis	Pimenta.
Pimenta acris*	Bay Laurel.
Pimpinella anisum	Anise.
Pimpinella Saxifraga	Saxifrage.
Pinus laricina*	Tamarac.
Pinus pendula*	Tamarac.
Pinus Strobus	White Pine.
Piper angustifolium	Matico.
Piper Cubeba	Cubeb.
Piper elongatum*	Matico.
Piper methysticum	Kava Kava.
Piper nigrum	Black Pepper.
Piscidia Erythrina	Jamaica Dogwood.
Plantago major	Plantain.
Poison hemlock	Conium.
Poison ivy	Poison Oak.

Polecatweed . . . . Skunk Cabbage.
Polemonium reptans . . . . Abscess root.
Polygala Senega. . . . . Senega. . . .
Polygonatum biflorum . . . . . Solomon's Seal.
Polygonatum giganteum . . . . Solomon's Seal.
Polygonatum officinale . . . . . Solomon's Seal.
Polygonum acre . . . . Water Pepper.
Polygonum Bistorta . . . . . . Bistort.
Polygonum Hydropiper* . . . . Water Pepper.
Polygonum hydropiperoides* . . Water Pepper.
Polygonum punctatum . . . . . Water Pepper.
Polymnia Uvedalia . . . . Bearsfoot.
Polyporus officinalis . . . . . . Agaric.
Polytrichum juniperinum . . . . Hair-cap Moss.
Populus balsamifera, v.candicans. Balm of Gilead.
Populus tremuloides . . . . . . White Poplar.
Potentilla Tormentilla . . . . . Tormentilla.
Prairie Pine-weed . . . . Button Snakeroot.
Prairie Pine-weed . . . . Rosinweed.
Premna taitensis . . . . Tonga.
Prince's Pine . . . . Pipsissewa.
Prinos verticillata . . . . Black Alder.
Prunus Persica . . . . Peach tree.
Prunus serotina . . . . Wild Cherry
Ptelea trifoliata . . . . Wafer Ash.
Pterocarpus Marsupium . . . Kino.
Pterocarpus santalinus . . . Red Saunders.
Pterocaulon pycnostachyum . . Indian, Black-root.
Pukeweed . . . . Lobelia.
Pulmonaria officinalis . . . . . Lungwort.
Punica Granatum . . . . . . Pomegranate.
Purging agaric . . . . Agaric.
Pussy Willow . . . . Black Willow.
Pycnanthemum montanum . . Mountain Mint.
Pyrola umbellata* . . . . Pipsissewa.
Pyrus Malus . . . . Apple-tree.
Quaking asp . . . . White Poplar.
Quassia excelsa* . . . . Quassia.
Quassia Simaruba* . . . . . . Simaruba.
Queen's delight . . . . Stillingia.
Queen's-root . . . . Stillingia.
Quercus alba . . . . White Oak bark.
Quercus lusitanica,
    var. *infectaria.* } Galls.
Quickens . . . . Couch-grass.
Raccoonberry . . . . Mandrake.
Ragwort . . . . Life-root.
Raphidophora vitiensis . . . . Tonga.
Rattleroot . . . . Black Cohosh.
Red Bud . . . . Judas-tree.
Red Centaury . . . . American Centaury.
Red Pepper . . . . Capsicum.
Red Puccoon . . . . Blood-root.

Red River Snakeroot	Serpentaria.
Red-root	Jersey Tea.
Rhamnus cathartica	Buckthorn berries.
Rhamnus Frangula	Buckthorn bark
Rheum officinale	Rhubarb.
Rheumatism-root	Twin-leaf.
Rhus Canadensis*	Rhus aromatica.
Rhus radicans	Poison Oak.
Richweed	Stone root.
Ricinus communis	Castor bean and leaves
Robin's Rye	Haircap Moss.
Rock Rose	Frostwort.
Rosa Gallica	Rose, Red.
Rottlera tinctoria	Kamala.
Roughroot	Button Snakeroot.
Roundleaved Dogwood	Green Osier.
Rubus Canadensis	Blackberry.
Rubus idaeus	Raspberry.
Rubus strigosus	Raspberry.
Rubus trivialis	Blackberry.
Rubus villosus	Blackberry.
Rudbeckia laciniata	Thimble-weed.
Rumex Acetosella	Sheep Sorrel.
Rumex crispus	Yellow Dock.
Rumex obtusifolius*	Yellow Dock.
Rumex sanguineus*	Yellow Dock.
Ruta graveolens	Rue.
Sabal serrulata*	Saw Palmetto.
Sabbatia angularis	American Centaury.
Sabbatia Elliottii	Quinine flower.
Sage Brush	Mountain Sage.
Saint John's-wort	Johnswort.
Salix alba	White Willow.
Salix nigra	Black Willow.
Salt-rheum weed	Balmony.
Salvia officinalis	Sage.
Sambucus Canadensis	Elder.
Sambucus nigra	European Elder.
Santalum album	Sandalwood.
Santalum rubrum	Red Saunders.
Santonica	Levant Wormseed.
Saponaria officinalis	Soapwort.
Sarothamnus Scoparius*	Broom.
Sarothamnus vulgaris*	Broom.
Sarracenia flava	Trumpet-plant.
Sarracenia purpurea	Pitcher-plant.
Sassafras officinale*	Sassafras.
Sassafras variifolium	Sassafras.
Satureia hortensis	Summer-savory.
Scopolia carniolica	Scopolia.
Scarletberry	Bittersweet.
Scilla maritima*	Squill.
Schoenocaulon officinale	Cevadilla.

Scrophularia Marylandica* . . . . Carpenter's Square.
Scrophularia nodosa . . . . . . Carpenter's Square.
Scutellaria canescens . . . . . . Scullcap.
Scutellaria lateriflora . . . . . Scullcap.
Seaweed . . . . Bladder wrack.
Seawrack . . . . Bladder wrack.
Senecio aureus . . . . Life-root.
Seneka Snakeroot . . . . Senega.
Serenoa serrulata . . . . Saw Palmetto.
Serratula spicata* . . . . Button Snakeroot.
Sesamum Indicum . . . . Benne.
Sesamum orientale* . . . . . . Benne.
Sevenbarks . . . . Hydrangea.
Sheep Laurel . . . . Mountain Laurel.
Silphium laciniatum . . . . . . Rosinweed.
Simaba Cedron . . . . Cedron.
Simaruba amara* . . . . Simaruba.
Simaruba excelsa* . . . . Quassia.
Skunkwweed . . . . Skunk Cabbage.
Small Spikenard . . . . American Sarsaparilla.
Smilax lanceolata . . . . Bamboo-brier.
Smilax medica . . . . Sarsaparilla.
Smilax officinalis . . . . Sarsaparilla.
Smilax ovata* . . . . Bamboo-brier root
Smilax papyraceæ . . . . Sarsaparilla.
Smilax Sarsaparilla . . . . Bamboo-brier.
Snakehead . . . . Balmony.
Snakeweed . . . . Bistort.
Snakeweed . . . . Serpentaria.
Snargel . . . . Serpentaria.
Solanum Carolinense . . . . . . Horse nettle.
Solanum Dulcamara . . . . . . Bittersweet.
Solanum paniculatum . . . . . Jurubeba.
Solidago odora . . . . Golden Rod.
Solidago Virgaurea . . . . . . Solidago.
Sophora tinctoria . . . . Wild Indigo.
Sorghum saccharatum . . . . . Broom Corn.
Southern Sarsaparilla . . . . . Bamboo-brier.
Spanish Chamomile . . . . . . Pellitory.
Spanish Fly . . . . Cantharides.
Spathyema foetida* . . . . . . Skunk Cabbage.
Spiraea tomentosa . . . . Hardhack.
Spotted Alder . . . . Witch Hazel.
Spotted Hemlock . . . . Water Hemlock.
Spotted Parsley . . . . Water Hemlock.
Squawbush . . . . Viburnum Opulus.
Squawmint . . . . Pennyroyal.
Squawroot . . . . Black Cohosh.
Squawroot . . . . Blue Cohosh.
Stachys Betonica . . . . Wood Betony.
Staphisagria macrocarpa* . . . Stavesacre.
Starwort . . . . False Unicorn.
Statice Limonium, v. Caroliniana. Marsh Rosemary.

Sterculia acuminata.	Kola.
Stigmata Maydis.	Corn Silk.
Stinging Nettle.	Nettle.
Stinkbush	Rhus aromatica.
Striped Alder.	Black Alder.
Strychnos Ignatia	Ignatia.
Strychnos Nux Vomica.	Nux Vomica.
Styrax Benzoin	Benzoin.
Swallowswort	Silkweed.
Swamp Alder.	Tag Alder.
Swamp Hellebore	Veratrum Viride.
Sweet Sumach.	Rhus aromatica.
Swertia Chirata	Chirata.
Sycocarpus Rusbyi.	Cocillana.
Symphytum officinale.	Comfrey.
Symplocarpus foetidus.	Skunk Cabbage.
Syzygium Jambolanum.	Jambul.
Tanacetum vulgare.	Tansy.
Taraxacum Dens-leonis*	Dandelion, Taraxacum.
Taraxacum officinale	Dandelion.
Terra Japonica.	Catechu.
Texas Sarsaparilla.	Yellow Parilla.
Texas Snakeroot.	Serpentaria.
Thea Chinensis.	Tea.
Thea sinensis	Tea.
Thorn Apple.	Stramonium.
Thoroughwort	Boneset.
Throatwort	Button Snakeroot.
Thuja (Thuya) occidentalis.	Arbor vitae.
Thymus Douglasii*	Yerba Buena.
Toluifera Balsamum.	Tolu.
Tonquin Bean	Tonka Bean.
Toothache tree.	Prickly Ash.
Toxicodendron crenatum*	Rhus aromatica.
Tree of Heaven	Ailanthus glandulosa.
Trifolium pratense.	Clover, Red.
Trilisa odoratissima.	Deer-tongue.
Trillium erectum.	Beth-root.
Trillium rhomboideum*.	Beth-root.
Triticum repens*	Couch-grass.
Trumpetweed	Queen of the Meadow
Tsuga Canadensis	Hemlock Spruce.
Tulip Poplar.	Tulip tree.
Turnera diffusa, v. aphrodisiaca.	Damiana.
Turnera microphylla*	Damiana.
Turtle bloom	Ba'mony.
Turtlehead.	Balmony.
Tussilago Farfara	Coltsfoot.
Umbellularia Californica	California Laurel.
Uncaria Gambier.	Catechu.
Unkum	Life-root.
Urginea maritima	Squill.
Urginea Scilla*	Squill.

Urtica dioica	Nettle root.
Valeriana angustifolia*	Valerian.
Valeriana sambucifolia*	Valerian.
Veratrum album*	White Hellebore.
Veratrum luteum*	False Unicorn.
Veratrum Sabadilla*	Cevadilla.
Veratrum viride	American Hellebore.
Verbascum Thapsus	Mullein.
Verbena hastata	Vervain, Blue.
Verbena paniculata*	Vervain, Blue.
Verbena urticaefolia	White Vervain.
Veronica Virginica	Culver's-root.
Viola tricolor	Pansy.
Virginia Creeper	American Ivy.
Virginia Snakeroot	Serpentaria.
Viscum flavescens*	Mistletoe.
Vitis quinquefolia*	American Ivy.
Wake Robin	Indian Turnip
Wake Robin	Beth-root.
Waterflag	Blue Flag.
Waxberry	Bayberry.
Wax myrtle	Bayberry.
White daisy	Ox-eye daisy.
White flag	Orris.
White saunders	Sandalwood.
White walnut	Butternut.
Whiteweed	Ox-eye daisy.
Whitewood	Tulip tree.
Wigandia Californica*	Yerba Santa.
Wild bryony	White bryony.
Wild cinnamon	Bay Laurel.
Wild cloves	Bay Laurel.
Wild hydrangea	Hydrangea.
Wild hyssop	Vervain, Blue.
Wild jessamine	Gelsemium.
Wild lemon	Mandrake.
Winter bloom	Witch Hazel.
Winter clover	Squaw Vine.
Wolfsbane	Aconite.
Woodbine	Gelsemium.
Woody nightshade	Bittersweet.
Xanthium Strumarium	Clotbur.
Xanthoxylum Americanum	Prickly Ash.
Xanthoxylum Carolinianum*	Prickly Ash.
Xanthoxylum Clava-Herculis	Prickly Ash.
Yellow cinchona	Cinchona (Calisaya).
Yellow gentian	American Columbo.
Yellow jessamine	Gelsemium.
Yellow poplar	Tulip tree.
Yellow puccoon	Golden Seal
Yellow root	Golden Seal.
Zea Mays	Corn-silk.
Zingiber officinale	Ginger.

INDEX.

INDEX.

ND - #0118 - 020223 - C0 - 229/152/16 - PB - 9780260120359 - Gloss Lamination